Duality Symmetry

Duality Symmetry

Special Issue Editor
Ivan Fernandez-Corbaton

MDPI • Basel • Beijing • Wuhan • Barcelona • Belgrade

Special Issue Editor
Ivan Fernandez-Corbaton
Karlsruhe Institute of
Technology (KIT),
Institute of Nanotechnology
Germany

Editorial Office
MDPI
St. Alban-Anlage 66
4052 Basel, Switzerland

This is a reprint of articles from the Special Issue published online in the open access journal *Symmetry* (ISSN 2073-8994) from 2019 to 2020 (available at: https://www.mdpi.com/journal/symmetry/special_issues/Duality_Symmetry).

For citation purposes, cite each article independently as indicated on the article page online and as indicated below:

LastName, A.A.; LastName, B.B.; LastName, C.C. Article Title. *Journal Name* **Year**, *Article Number*, Page Range.

ISBN 978-3-03936-569-2 (Hbk)
ISBN 978-3-03936-570-8 (PDF)

© 2020 by the authors. Articles in this book are Open Access and distributed under the Creative Commons Attribution (CC BY) license, which allows users to download, copy and build upon published articles, as long as the author and publisher are properly credited, which ensures maximum dissemination and a wider impact of our publications.

The book as a whole is distributed by MDPI under the terms and conditions of the Creative Commons license CC BY-NC-ND.

Contents

About the Special Issue Editor vii

Preface to "Duality Symmetry" ix

Edvard T. Musaev
U-Dualities in Type II and M-Theory: A Covariant Approach
Reprinted from: *Symmetry* **2019**, *11*, 993, doi:10.3390/sym11080993 1

Alberto Blasi and Nicola Maggiore
Topologically Protected Duality on The Boundary of Maxwell-BF Theory
Reprinted from: *Symmetry* **2019**, *11*, 921, doi:10.3390/sym11070921 32

Ivan Fernandez-Corbaton
A Conformally Invariant Derivation of Average Electromagnetic Helicity
Reprinted from: *Symmetry* **2019**, *11*, 1427, doi:10.3390/sym11111427 44

Philipp Gutsche, Xavier Garcia-Santiago, Philipp-Immanuel Schneider, Kevin M. McPeak, Manuel Nieto-Vesperinas and Sven Burger
Role of Geometric Shape in Chiral Optics
Reprinted from: *Symmetry* **2020**, *12*, 158, doi:10.3390/sym12010158 56

Yosuke Nakata, Yoshiro Urade, and Toshihiro Nakanishi
Geometric Structure behind Duality and Manifestation of Self-Duality from Electrical Circuits to Metamaterials
Reprinted from: *Symmetry* **2019**, *11*, 1336, doi:10.3390/sym11111336 64

Lisa V. Poulikakos, Jennifer A. Dionne and Aitzol García-Etxarri
Optical Helicity and Optical Chirality in Free Space and in the Presence of Matter
Reprinted from: *Symmetry* **2019**, *11*, 1113, doi:10.3390/sym11091113 117

About the Special Issue Editor

Ivan Fernandez-Corbaton was awarded a degree in Electrical Engineering from the Polytechnic University of Catalonia (Barcelona, Spain) in 1998, and an MSc in Mobile Communications from the Eurecom Institute (Sophia Antipolis, France). From 1998 to 2010 he worked as a research engineer, mostly in the design of signal processing algorithms for cellphone chips. Ivan has 40 patents from his engineering period. He moved back into academia as a PhD student in 2010. In 2014 he obtained his PhD in physics from Macquarie University (Sydney, Australia). Since 2014, has Ivan worked in the Karlsruhe Institute of Technology (Karlsruhe, Germany). In his physics research, he studies light–matter interactions by means of symmetries and conservation laws. Chirality in electromagnetic interactions is one of Ivan's main research themes, and includes the search for improved theoretical insights as well as their subsequent practical application, for example, for the enhanced sensing of chiral molecules. The symmetry conditions for the suppression of the backscattering of light, and its engineering in realistic systems like solar cells, has also been a recurring theme in Ivan's research. Lately, Ivan is becoming increasingly interested in the quantification of symmetry breaking.

Preface to "Duality Symmetry"

I would like to warmly thank all the authors and reviewers of this Special Issue for their contribution to the process. This Special Issue is composed of four Articles, one Review and one Perspective. The context of the contributions ranges from string theory to applied nanophotonics, which, as anticipated, shows that duality symmetries, and electromagnetic duality symmetry in particular, are useful in a wide variety of physics fields, both theoretical and applied. Moreover, several of the contributions show how the use of symmetry arguments and the quantification of symmetry breaking can successfully guide our theoretical understanding and provide us with guidelines for system design. I fervently anticipate the practicality and future outcomes of such research avenues.

Ivan Fernandez-Corbaton
Special Issue Editor

Review

U-Dualities in Type II and M-Theory: A Covariant Approach

Edvard T. Musaev

Moscow Institute of Physics and Technology, Institutskii per. 9, 141700 Dolgoprudny, Russia; musaev.et@phystech.edu

Received: 12 June 2019; Accepted: 23 July 2019; Published: 3 August 2019

Abstract: In this review, a short description of exceptional field theory and its application is presented. Exceptional field theories provide a U-duality covariant description of supergravity theories, allowing addressing relevant phenomena, such as non-geometricity. Some applications of the formalism are briefly described.

Keywords: T-duality; exotic branes; non-geometric backgrounds; exceptional field theory

1. Introduction

The notion of symmetry has been one of the most important drivers for theoretical physics in recent decades. Recently, of most interest have been duality symmetries relating different theories or different regimes of the same theory. The most well-known example of the former is the renowned AdS/CFT correspondence of Maldacena [1], and more generally gauge-gravity duality. This correspondence relates string theory on $AdS_5 \times \mathbb{S}^5$ at small coupling and the quantum $\mathcal{N}=4$ SYM at strong coupling. This is based on equivalence between two descriptions of the same near-horizon region of the D3-brane in terms of the open and closed strings, and allows addressing phenomena in CFT at strong coupling in terms of gravitational degrees of freedom at weak coupling.

String theory also possesses symmetries, called T- and S-dualities, which relate different regimes of the same theory. These allowed unifying five string theories, Type IIA/B, Type I, heterotic $O(32)$ and $E_8 \times E_8$, into a duality web and to understand all these as different regimes of a single theory, usually addressed as M-theory [2,3]. T-duality is a perturbative symmetry of the fundamental string and is the oldest known duality in string theory. It manifests itself in the amazing fact that Type IIA and Type IIB string theories compactified on a 1-torus $\mathbb{S}^1$ are equivalent at quantum level. Most transparently this can be seen when looking at the mass spectrum of the closed string on a background with one circular direction of radius R

$$M^2 = p^2 + \frac{n^2}{R^2} + \frac{m^2 R^2}{\alpha'^2} + 2(N + \tilde{N} - 2). \tag{1}$$

Here n is the number of discrete momenta of the string along the circular direction and m is the number of windings of the string around the cycle, N and $\tilde{N}$ correspond to the numbers of higher level left and right modes on the string. One immediately notices that the mass spectrum is symmetric under the following replacement

$$\begin{aligned} R &\longleftrightarrow \frac{\alpha'}{R}, \\ m &\longleftrightarrow n. \end{aligned} \tag{2}$$

The symmetry relates backgrounds with radii R and α'/R upon switching string momentum and winding modes. More generally T-duality mixes metric and 2-form gauge field degrees of freedom, thus potentially completely messing up structure of space-time. In particular, the symmetry allows consistently defining string dynamics on such backgrounds, given by explicit metric $G_{\mu\nu}$ and the Kalb-Ramond field $B_{\mu\nu}$ that cannot be consistently described in terms of manifolds. Instead, these are related to as T-folds, which are defined as a set of patches glued by T-duality transformations, rather than diffeomorphisms [4]. Such backgrounds are called non-geometric and are of huge interest for cosmological model building as string vacua potentially capable of completely stabilizing scalar moduli ending up with a dS-like space with small cosmological constant.

T-duality is a perturbative symmetry of string theory seen already in the mass spectrum of string excitations. S-duality of Type IIB string theory provides an example of non-perturbative string symmetry. This relates strong and weak coupled regimes of the theory, and in addition relates heterotic SO(32) and Type I strings. The net of dualities between five string theories allows understanding them as different approximations to a single 11-dimensional theory called M-theory. The 11th direction arises in the strong coupling limit of Type IIA theory.

M-theory describes dynamics of M2 and M5 branes, which are fundamental 2 and 5 dimensional objects interacting with 3-form and 6-form gauge potentials dual to each other [3]. Low-energy limit of the theory is captured by 11-dimensional maximal supergravity, whose algebra of central charges is nicely interpreted in terms of M2 and M5 brane charges [3]. M-theory compactified on a circle $\mathbb{S}^1$ gives Type IIA theory with the fundamental string arising from wrapped M2 brane. Compactifying M-theory on a 2-torus one recovers either Type IIA or Type IIB depending on the reduction scheme, which is the fundamental precursor for T-duality symmetry between these theory. In addition, modular symmetry of the torus gives rise to S-duality symmetry of Type IIB as will be explained in more details below. Together, T- and S-duality of the string combine into a set of U-duality symmetries, which appears to be a powerful tool for investigating properties and the internal structure of M-theory. This review is focused on duality symmetries of string and M-theory and in particular at approaches to supergravity covariant with respect to these symmetries.

1.1. Dualities in String Theory

S-duality is a hidden symmetry of the Type IIB string theory which relates strong and weak coupling regimes. On the field theory level this can be recovered by inspecting spectrum of massless modes of the Type IIB string, which includes

$$g, \quad \phi, \quad B_{(2)}, \quad C_{(0)}, \quad C_{(2)}, \quad C_{(4)}, \tag{3}$$

where g is the metric, ϕ is the scalar field called the dilaton and the 4-form gauge field $C_{(4)}$ is defined to have self-dual field strength. With the action of the S-duality group $SL(2,\mathbb{R})$ the fields drop into irreducible representations with g being a scalar, the 2-forms $B_{(2)}, C_{(2)}$ combine into a doubled and the scalar fields $\phi, C_{(0)}$ combine into the so-called axio-dilaton defined as

$$\tau = C_{(0)} + ie^{-\phi}. \tag{4}$$

Axio-dilaton transforms non-linearly under S-duality

$$\tau' = \frac{a\tau + b}{c\tau + d}, \quad \begin{bmatrix} a & b \\ c & d \end{bmatrix} \in SL(2,\mathbb{R}). \tag{5}$$

The existence of this hidden symmetry of the Type IIB massless string spectrum suggests that the supergravity action can be rewritten in an $SL(2,\mathbb{R})$-covariant form, which is indeed possible and results in the following (see e.g., [5]):

$$\begin{aligned} S_{IIB} = -\frac{1}{2} \int d^{10}x \Big(& R - \frac{1}{4}\mathrm{Tr}(\partial M \partial M^{-1}) + \frac{3}{4} H_{\mu\nu\rho}{}^I M_{IJ} H^{\mu\nu\rho J} \\ & + \frac{5}{6} F_{\mu\nu\rho\sigma} F^{\mu\nu\rho\sigma} + \frac{1}{96\sqrt{-g}} \epsilon_{IJ} C_{(4)} \wedge H_{(3)}{}^I \wedge H_{(3)}{}^J \Big), \end{aligned} \tag{6}$$

where the indices $I, J, \cdots = 1, 2$ label the fundamental **2** representation of SL(2) and

$$M_{IJ} = \frac{1}{\Im\tau} \begin{bmatrix} |\tau|^2 & -\Re\tau \\ -\Re\tau & 1 \end{bmatrix}, \tag{7}$$

$$H_{\mu\nu\rho}{}^I = \partial_{[\mu} B_{\nu\rho]}{}^I, \quad F_{\mu\nu\rho\sigma\kappa} = \partial_{[\mu} C_{\nu\rho\sigma\kappa]} + \frac{3}{4}\epsilon_{IJ} B_{[\mu\nu}{}^I \partial_\rho B_{\sigma\kappa]}{}^J.$$

One may notice that the transformation (5) has the form that of the transformation of complex structure of a 2-torus. This observation leads to the geometrical picture which is behind F-theory, a 12-dimensional field theory, whose compactification on a 2-torus gives an orientifold reduction of Type IIB theory (for more details see e.g., [6,7]). Freedom in definition of a complex structure on the 2-torus of F-theory is equivalent to the S-duality symmetry of the 10-dimensional theory. Such geometric interpretation of a symmetry of a theory goes along the line of the old Kaluza-Klein idea, where the U(1) gauge symmetry of Maxwell theory is lifted into reparametrisations of a small circle (1-torus) of a 5-dimensional gravitational theory with no electromagnetic degrees of freedom. In this short review we highlight basic features and list some applications of the so-called Doubled (Exceptional) Field Theory, which does the same to T(U)-dualities of string (M-)theory, i.e., provides a geometric interpretation of the duality symmetries in terms of geometry of an especially constructed higher dimensional space.

T-duality is a hidden symmetry of the 2-dimensional non-linear sigma model (string theory) which relates the theory on a torus with radius R_x of a given direction x and the same theory on a torus with radius α'/R_x of the same direction. Under T-duality transformation of the string background, given by the metric, Kalb-Ramond 2-form field, the dilaton and the RR fields, partition function of the string does not change. Consider the action for the closed string on a background with one circular direction in the conformal gauge and adopt the light cone world-sheet coordinates $\sigma_\pm$

$$\begin{aligned} S_1[\theta] &= \int d^2\sigma \, (G+B)_{\mu\nu} \partial_+ X^\mu \partial_- X^\nu \\ &= \int d^2\sigma \left(G_{\theta\theta} \partial_+ \theta \partial_- \theta + E_{\hat{a}\theta} \partial_+ X^{\hat{a}} \partial_- \theta + E_{\theta \hat{a}} \partial_+ \theta \partial_- X^{\hat{a}} + E_{\hat{a}\hat{\beta}} \partial_+ X^{\hat{a}} \partial_- X^{\hat{\beta}} \right). \end{aligned} \tag{8}$$

Here $\mu, \nu = 0, \ldots, 9$ label all ten space-time directions, $\theta = \theta(\sigma_\pm)$ parametrizes the compact direction and $\hat{\alpha}, \hat{\beta} = 0, \ldots, 8$ parametrize the rest. For the background fields we define $E = G + B$. To see that the action above is invariant under replacing the circle $\mathbb{S}^1_\theta$ of radius R by a circle $\mathbb{S}^1_\lambda$ of the inverse radius $1/R$ parametrized by λ, one used the global symmetry $\theta \to \theta + \zeta$ and turns it into a local one. The gauging is performed by introducing a world-volume 1-form gauge field $A = A_+ d\sigma^+ + A_- d\sigma^-$ and replacing normal derivatives by covariant e $d\theta \to D\theta = d\theta + A$. To turn back to the correct counting of the degrees of freedom in the theory one must introduce a Lagrange multiplier to restrict the gauge field to pure gauge. Hence, one arrives at the following action

$$S_2[\theta, \lambda, A] = S_1[d\theta \to D\theta] + \int \lambda F, \tag{9}$$

where $F = dA$ is the field strength for the gauge field A. Integrating the Lagrange multiplier λ out of the partition function one return back to the action $S_1[\theta]$. Alternatively, integrating out vector degrees of freedom A one arrives at (for more details see [8])

$$S_2[\lambda] = \int d^2\sigma \left(G'_{\lambda\lambda} \partial_+ \lambda \partial_- \lambda + E'_{\hat{a}\lambda} \partial_+ X^{\hat{a}} \partial_- \lambda + E'_{\lambda\hat{a}} \partial_+ \lambda \partial_- X^{\hat{a}} + E'_{\hat{a}\hat{\beta}} \partial_+ X^{\hat{a}} \partial_- X^{\hat{\beta}} \right), \tag{10}$$

with the new background $E' = G' + B'$ defined by the so-called Buscher rules

$$
\begin{aligned}
G'_{\lambda\lambda} &= \frac{1}{G_{\theta\theta}}, \\
E'_{\lambda\hat{\alpha}} &= \frac{1}{G_{\theta\theta}} E_{\theta\hat{\alpha}}, \\
E'_{\hat{\alpha}\lambda} &= -\frac{1}{G_{\theta\theta}} E_{\hat{\alpha}\theta}, \\
E'_{\hat{\alpha}\hat{\beta}} &= E_{\hat{\alpha}\hat{\beta}} - E_{\hat{\alpha}\theta} \frac{1}{G_{\theta\theta}} E_{\theta\hat{\beta}}.
\end{aligned}
\tag{11}
$$

Since the partition function did not change during the above procedure, the physics of the string on two background related by the T-duality transformation (11) is the same. Taking into account the transformation of measure of the functional integral one supplements the above rules by the following transformation of the dilaton:

$$
\varphi' - \frac{1}{4} \ln \det G' = \varphi - \frac{1}{4} \ln \det G.
\tag{12}
$$

In the more general case of the string on a background with d compact isometric directions the group of T-duality transformations can be shown to be $O(d,d;\mathbb{Z})$. The most convenient tool for that is the so-called Duff's procedure [9,10] which exploits hidden symmetry between equations of motion and Bianchi identities following from the action for the non-linear sigma model

$$
S = \int d^2\sigma \left(\sqrt{-h}\, h^{ab} G_{\mu\nu} + \epsilon^{ab} B_{\mu\nu} \right) \partial_a X^\mu \partial_b X^\nu
\tag{13}
$$

For the metric and the B-field equations of motion and Bianchi identities can be rewritten in explicitly $O(d,d;\mathbb{Z})$-covariant form

$$
\eta_{MN} \tilde{\Phi}^{iN} = \mathcal{H}_{MN} \Phi^{iN}.
\tag{14}
$$

Here we define combinations

$$
\tilde{\Phi}^{iM} = \begin{bmatrix} \epsilon^{ab} \partial_b X^\mu \\ \epsilon^{ab} \partial_b Y_\mu \end{bmatrix}, \quad \Phi^{iM} = \begin{bmatrix} \sqrt{-h}\, h^{ab} \partial_b X^\mu \\ \sqrt{-h}\, h^{ab} \partial_b Y_\mu \end{bmatrix}
\tag{15}
$$

of derivatives of the normal coordinates X^μ and the dual coordinates Y_μ. The equation (14) can be considered to be self-duality constraints, which remove half of the fields of the full doubled set $\mathbb{X}^M = (X^\mu, Y_\mu)$. The matrix $\mathcal{H}_{MN}$ parametrizes the background fields in a T-duality covariant manner

$$
\mathcal{H}_{MN} = \begin{bmatrix} G_{\mu\nu} - B_{\mu\rho} B^\rho{}_\nu & -B_\mu{}^\nu \\ B^\mu{}_\nu & G^{\mu\nu} \end{bmatrix} \in \frac{O(d,d)}{O(d) \times O(d)}.
\tag{16}
$$

The invariant tensor of the $O(d,d)$ group η_{MN} is taken in the block-skew-diagonal form

$$
\eta_{MN} = \begin{bmatrix} 0 & \delta^\mu_\nu \\ \delta_\nu^\mu & 0 \end{bmatrix}.
\tag{17}
$$

At the level of string theory T-duality is a proper symmetry of the theory, which does not change physics upon a transformation. When reducing to the low-energy dynamics governed by 10-dimensional half-maximal supergravity, T-duality turns into a solution-generating symmetry, as it transforms a given string theory background into another one. In this case, the symmetry group

$O(d, d; \mathbb{R})$ is defined over rational numbers rather than only integers. In what follows, we will always denote this group simply by $O(d, d)$, and add $\mathbb{Z}$ explicitly when needed.

1.2. U-duality in Maximal Supergravity

When combined into the web of dualities five string theories become a single 11-dimensional M-theory, encoded in dynamics of M2 and M5 branes. T- and S-duality symmetries lift into U-dualities of the membranes; however, these are much more complicated for a sigma model analysis. In the seminal paper by Cremmer, Julia, Lu and Pope [11,12] it has been shown that 11-dimensional supergravity compactified on a d-torus $\mathbb{T}^d$ possesses hidden symmetry $E_{d(d)}$. This is reflected in the fact that all bosonic field of the reduced theory can be collected into irreducible representations of the U-duality group, while fermionic fields transform under maximal compact subgroup of $E_{d(d)}$. Here the notation $d(d)$ means that one takes maximal real subgroup of complexification of the group E_d.

The most transparent way to see the symmetry is to analyse spectrum of fields in the lower dimensional theory obtained by reduction of the 11-dimensional fields $G_{\hat{\mu}\hat{\nu}}, C_{\hat{\mu}\hat{\nu}\hat{\rho}}$ say to 4 dimensions.

As it is shown in Figure 1 the resulting fields can be collected into the vector fields $\mathcal{A}_\mu{}^M$ transforming in the $\mathbf{56}$ of E_7, scalar coset $M_{MN} \in E_{7(7)}/SU(8)$, E_7 scalars $G_{\mu\nu}$ and a constant $C_{\mu\nu\rho}$. Vector degrees of freedom $(A_\mu{}^m, A_{\mu mn})$ coming from the metric and the 3-form field correspond 28 electric gauge potentials. To compose the $\mathbf{56}$ irrep of E_7, which is representation of the lowest dimension, one adds magnetic potentials $(\tilde{A}_{\mu m}, \tilde{A}_\mu{}^{mn})$ and imposes self-duality condition on the U-duality covariant field strength

$$\mathcal{F}_{\mu\nu}{}^M = \frac{i}{2}\epsilon_{\mu\nu\rho\sigma}\Omega^{MN}M_{NK}\mathcal{F}^{\rho\sigma K}. \tag{18}$$

Here Ω^{MN} is the symmetric invariant tensor of E_7 and M_{MN} is the scalar matrix which parametrizes the coset space $E_{7(7)}/SU(8)$ and the antisymmetric tensor is usually chosen to be $\epsilon_{0123} = i$. The field strength is defined as usual as $\mathcal{F}_{\mu\nu}{}^M = 2\partial_{[\mu}\mathcal{A}_{\nu]}{}^M$. As it has been explicitly shown in detail in [11], to end up with irreducible representations of the U-duality group one must dualize all tensor fields to the lowest possible rank. For the 2-form field $B_{\mu\nu m}$ one constructs gauge invariant 3-form field strength, whose Bianchi identities and equations of motion can be swapped by Hodge dualization to a 1-form field strength. This is associated with scalar fields. To keep covariancy and to recover tensor hierarchy one must add the same amount of 2-form fields and impose duality condition.

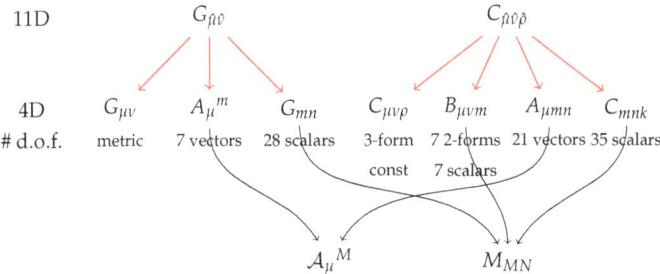

Figure 1. Reduction of 11D fields into four dimensions, dualization into the lowest possible rank tensor and recollection into the E_7 multiplet in the $\mathbf{56}$ and the coset space $E_{7(7)}/SU(8)$. Both magnetic and electric vectors potentials are included in the counting (see the text).

Finally, equations of motion for the 3-form field in 4 dimensions imply that the field strength is constant. It appears that this constant is not a scalar under the U-duality group, and one must either set it to zero or to turn on all constant belonging to its representation. This can be done consistently in embedding tensor formalism, which results in a deformed theory with non-abelian gauge symmetry [13].

In conventional supergravity U-duality symmetry is a global symmetry of the theory similar to the U(1) duality symmetry of equations of motion of electrodynamics. In exceptional field theories this symmetry receives a geometric interpretation as a symmetry of the underlying space-time. To carry a proper representation of a U-duality group E_d space-time coordinates must be extended following a special rule based on winding of branes, both standard and exotic. In Section 2.1 the construction and local symmetries of the extended space will be explained in more details, in Section 2.3 field spectrum and action of exceptional field theories will be reviewed for a general U-duality group and for $E_{7(7)}$ as an explicit example. We will show that this theory reproduces the full 11D supergravity, Type IIA/B, D = 4 gauged and ungauged supergravities depending on the choice of solution of a special constraint called section condition. In Section 3.1 non-conventional solutions of this condition will be shown to lead to non-geometric backgrounds, i.e., such field configurations which are not globally or even locally well defined in terms of differential geometry. Finally, in Section 3.2 we consider exotic branes as T(U)-duality partners of the standard branes of string and M-theory, and review their description in terms of extended space and non-geometry in exceptional field theories.

2. Duality-Covariant Field Theories

2.1. Local Symmetries of Extended Space

Duality symmetries appear in toroidal reductions of supergravity and combine geometric symmetries of the torus (diffeomorphisms), gauge transformations and actual duality transformations mixing space-time and gauge degrees of freedom. To proceed with construction of a duality covariant theory, one understands the group of duality symmetries $O(d,d)$ or $E_{d(d)}$ more fundamentally as descending from geometrical structure of the underlying space. Since d coordinates of the d-torus do not fit into an irreducible representation of the duality group, one has to consider an extended space.

In the previous section, we saw that T-duality symmetries relate winding and momentum modes of the string, and for the string on a torus $\mathbb{T}^d$ the mass spectrum can be covariantly written as

$$M^2 = p^2 + \mathcal{H}_{MN}\mathcal{P}^M\mathcal{P}^N + (N + \tilde{N} - 2), \tag{19}$$

where $\mathcal{H}_{MN}$ is the generalized metric and $\mathcal{P}^M$ combines the momentum p_m and winding w^m of the string in an $O(d,d)$-covariant vector

$$\mathcal{P}^M = \begin{bmatrix} p_m \\ w^m \end{bmatrix}. \tag{20}$$

The first term p^2 contains momenta in "external" non-toroidal directions $p^2 = \eta_{\mu\nu}p^\mu p^\nu$. Recalling Duff's procedure, that operates with normal X^m and dual Y_m scalar fields on the world-volume of the string, it is natural to consider backgrounds, not necessarily toroidal that depend on the full set of T-duality covariant coordinates $\mathbb{X}^M = (x^m, \tilde{x}_m)$. Note that the generalized metric in the mass formula is constant since the background is toroidal, this is relaxed in exceptional field theory and the toroidal case is understood as the most symmetric background preserving all duality symmetries.

At the level of the non-linear sigma model extra degrees of freedom encoded by the dual scalar fields Y_m are removed by the self-duality condition (14). Similarly, T-covariant field theory with fields depending on the doubled amount of coordinates must be augmented by a constraint that reduces the number of space-time direction. Apart from reference to the sigma model, this is necessary for further supersymmetric completion of the theory to avoid higher spin fields, which normally appear in supersymmetric theories in dimensions higher than 11. The condition naturally follows from construction of local diffeomorphism symmetry on the extended space parametrized by the coordinates $\mathbb{X}^M$ which is defined by the so-called generalized Lie derivative

$$\mathcal{L}_\Lambda V^M = \Lambda^N \partial_N V^M - \Lambda^M \partial_M V^N + Y^{MN}_{KL} \partial_N \Lambda^K V^L, \tag{21}$$

where $V^M = (v^m, \omega_m)$ is a generalized vector combining a GL(d) vector v^n and a 1-form ω_m, the same for the transformation parameter $\Lambda^M = (\lambda^m, \tilde{\lambda}_m)$. The tensor Y^{MN}_{KL} encoding deformation of the generalized Lie derivative away from the conventional GL(d) Lie derivative is subject to constraints following from consistency of algebra of transformation $\delta_\Lambda V^M = \mathcal{L}_\Lambda V^M$. These constraints have been analysed in [14] and can be summarized as

$$\begin{aligned} Y^{(MN}_{KL} Y^{R)L}_{PQ} - Y^{(MN}_{PQ} \delta^{R)}_K &= 0, \text{ for } d \leq 5, \\ Y^{MN}_{KL} &= -\alpha_d P_K{}^M{}_L{}^N + \beta_d \delta^M_K \delta^N_L + \delta^M_L \delta^N_K, \\ Y^{MA}_{KB} Y^{BN}_{AL} &= (2 - \alpha_d) Y^{MN}_{KL} + (D\beta_d + \alpha_d) \beta_d \delta^M_K \delta^N_L + (\alpha_d - 1) \delta^M_L \delta^N_K. \end{aligned} \tag{22}$$

Here d is the number of compact dimensions and $P_A{}^B{}_C{}^D$ is the projector on the adjoint representation of the corresponding duality group. It is defined as $P_A{}^B{}_C{}^D P_D{}^C{}_K{}^L = P_A{}^B{}_K{}^L$ and $P_A{}^B{}_B{}^A = \dim(\text{adj})$. The coefficients α_d and β_d depend on the duality group and for the cases in question take numerical values $(\alpha_4, \beta_4) = (3, \frac{1}{5})$, $(\alpha_5, \beta_5) = (4, \frac{1}{4})$, $(\alpha_6, \beta_6) = (6, \frac{1}{3})$. These conditions imply that the Y-tensor must be constructed from invariant tensors of the corresponding T- and U-duality groups (see Table 1).

Table 1. Y-tensor for different T(U)-duality groups for string and M-theory on a d-torus. Here the Greek indices $\alpha, \beta, \gamma = 1, \ldots, 5$ label the representation **5** of $SL(5)$ and the index i labels the **10** of $SO(5,5)$, n denotes dimension of the representation generalized vectors of the theory transform in.

Duality Group	The Y-tensor	Dimension of Extended Space
O(d,d)	$Y^{MN}_{KL} = \eta^{MN} \eta_{KL}$	$n = 2d$
SL(5)	$Y^{MN}_{KL} = \epsilon^{\alpha MN} \epsilon_{\alpha KL}$	$n = 10$
SO(5,5)	$Y^{MN}_{KL} = \frac{1}{2}(\gamma^i)^{MN}(\gamma_i)_{KL}$	$n = 16$
$E_{6(6)}$	$Y^{MN}_{KL} = 10\, d^{MNR} d_{KLR}$	$n = 27$
$E_{7(7)}$	$Y^{MN}_{KL} = 12\, c^{MN}{}_{KL} + \delta^{(M}_K \delta^{N)}_L + \frac{1}{2} \epsilon^{MN} \epsilon_{KL}$	$n = 56$

The above still does not guarantee the algebra of transformations $\delta_\Lambda V^M$ is closed. Indeed, one writes

$$\mathcal{L}_{\Lambda_1} \mathcal{L}_{\Lambda_2} V^M - \mathcal{L}_{\Lambda_2} \mathcal{L}_{\Lambda_1} V^M = \mathcal{L}_{[\Lambda_1, \Lambda_2]_C} V^M + F^M_0, \tag{23}$$
$$F^M_0 =$$

where the bracket $[\Lambda_1, \Lambda_2]_C = \mathcal{L}_{L_1} \Lambda_2 - \mathcal{L}_{L_2} \Lambda_1$ is a generalisation of the Courant bracket of the Hitchin's generalised geometry. The obstruction F^M_0 =for the algebra to close is proportional to terms of the type $\eta^{MN} \partial_M \Phi_1 \partial_N \Phi_2$, hence one naturally imposes the so-called section constraint

$$\eta^{MN} \partial_M \bullet \partial_N \bullet = 0 \tag{24}$$

where bullets stand for any combination of any fields. Similarly, one shows that the very same condition ensures satisfaction of the Jacobi identity for δ_Λ.

The most natural and transparent solution of the section constraint is $\tilde{\partial}^m \bullet = 0$, which is simply the condition that nothing depends on $\tilde{x}_m$. This drops the generalized Lie derivative back to the conventional undoubled space-time and splits it into the usual Lie derivative and gauge transformations. More generally one can solve the section constraint by dropping dependence on all $\tilde{x}_m$ apart a given $\tilde{x}_d$, and drop in addition dependence on the corresponding normal coordinate x^d. Now one can use the set $\{x^1, \ldots, x^{d-1}, \tilde{x}_d\}$ to measure distances, and hence these will correspond to geometric coordinates of the new frame. These two frames are related by a T-duality transformation along the direction d

$$T_d : x^d \longleftrightarrow \tilde{x}_d. \quad (25)$$

In the next subsection we construct Exceptional Field Theories, which do not distinguish between such frames, hence providing a local T(U)-duality covariant approach to supergravity.

2.2. Winding Modes and Exotic Branes

Before proceeding with the field theory construction it is suggestive to follow the logic of counting of winding modes of M-branes in M-theory in details. In contrast to the string, where the winding mode is always parametrized by a 1-form w_m irrespective of the number of compact directions, winding modes of branes follows more complicated pattern. To start with, one notices that winding modes of a p-brane can be parametrized by a p-form. Spectrum of M-theory contains M2, M5-branes, KK6-monopole and various additional (exotic) branes, whose counting will be useful for U-duality groups $E_{8(8)}$ and larger. Table 2 lists irreducible representations of U-duality groups for each dimension d of compact torus, governing transformation of extended coordinates $\mathbb{X}^M$ and the corresponding generalized momentum $\mathcal{P}_M$. The normal geometric coordinates correspond to the usual momentum P of a state. Windings of M2 and M5 branes are given by 2- and 5-forms respectively and give C_d^2 and C_d^5 number of winding states, where C_m^n is the binomial coefficient. Hence, the M5 brane contributes only starting from dimension $d = 5$, since it simply cannot wrap spaces of lower dimensions.

Table 2. Counting of winding modes of branes of M-theory on a background of the form $M_4 \times T^d$ with M_4 being a four-dimensional manifold. The first column contains dimensions of the compact torus, the second column lists the corresponding U-duality group, and the last column lists representations of G under which coordinates of the extended space transform.

d	G	P	M2	M5	KK6	5^3	2^6	$0^{(1,7)}$	$\mathcal{R}_\mathbb{X}$
2	SL(2)	2	1	-	-	-	-	-	3
3	SL(3)×SL(2)	3	3	-	-	-	-	-	(3,2)
4	SL(5)	4	6	-	-	-	-	-	10
5	SO(5,5)	5	10	1	-	-	-	-	16_s
6	$E_{6(6)}$	6	15	6	-	-	-	-	27
7	$E_{7(7)}$	7	21	21	7	-	-	-	56
8	$E_{8(8)}$	8	28	56	56	56	28	8	248

The Kaluza-Klein monopole KK6 is an object with 6 + 1-dimensional worldvolume and one Taub-NUT direction corresponding to the Hopf cycle. This is magnetic dual of the graviton. Hence, it windings are represented by a mixed-symmetry tensor $z_{(7,1)}$, which is a 7-form taking values in 1-forms and traceless. In components this is represented by the following tensor

$$z_{a_1 \ldots a_7, b}, \quad (26)$$

where b must be equal to one of a_i's for non-vanishing components. For $d = 7$ winding direction this amounts in 7 independent winding modes, while for the $E_{8(8)}$ case is is suggestive to contract $z_{a_1 \ldots a_7, b}$ with Levi-Civita tensor as

$$z_b{}^a = z_{a_1 \ldots a_7, b}\, \epsilon^{a_1 \ldots a_7 a}. \quad (27)$$

In total $z_a{}^b$ has $8 \times 8 = 64$ components and the condition $z_a{}^a\big|_{\text{no sum}} = 0$ removes 8 more leaving only 56.

The important observation here is that the total number of momentum and winding modes for a d-torus with $d < 8$ sums up to dimension of an irreducible representation of the corresponding U-duality symmetry group. For $d = 8$ this apparently does not work, as summing winding modes of all branes up to the KK6 one obtains 148, while dimension of the smallest irrep is 248. To cure the result one first recalls that the spectrum of both string and M-theory contains exotic branes in addition to the normal (standard) branes. These are T(U)-duality partners of the normal branes and can be understood as sources of non-geometric backgrounds. Such backgrounds cannot be defined in terms of manifolds, instead these are described in terms of T(U)-folds [4,15,16], whose patches are glued by T(U)-duality transformations. At the level of field configurations this is realized as a monodromy when going around the point the exotic brane is placed [17].

It is convenient to label exotic branes in the same way as states of the 3D maximal supergravity are classified [3,17]. Hence, for any brane of string theory one adopts the notation $b_n^{(c_r...c_1)}$, where $b + 1$ gives the number of world-volume directions, c_i denote the number of special (quadratic, cubic etc.) directions and n gives the power of the string coupling constant g_s in tension of the brane. Such brane completely wrapped on a torus would have tension given by

$$M\left[b_n^{(c_r...c_1)}\right] = \frac{R_{i_1}\ldots R_{i_b} R_{j_1}^2 \ldots R_{j_{c_1}}^2 \ldots}{g_s^n l_s^{b+2c_1+3c_2+\ldots+(r+1)c_r+1}}, \tag{28}$$

where R_i denote radius of the i-th toroidal direction and l_s is the string length. For example, for the NS5-brane, which has 6 world-volume directions, no special circles and whose tension scales as g_s^{-2}, one would use the notation $NS5 = 5_2^0 \equiv 5_2$. Kaluza-Klein monopole is denoted as $KK5 = 5_2^1$ and has 6 worldvolume directions, one special circle corresponding to the Hopf fibre and its tension scales as g_s^{-2}. In these notations duality symmetries of string theory act on such states as follows

$$T_x: \quad R_x \to \frac{l_s^2}{R_x}, \quad g_s \to \frac{l_s}{R_x} g_s; \quad S: \quad g_s \to \frac{1}{g_s}, \quad l_s \to g_s^{\frac{1}{2}} l_s. \tag{29}$$

The well-known example of a T-duality orbit containing exotic branes has been investigated in [18] (see also [19] for a review) and reads

$$5_2^0 \to 5_2^1 \to 5_2^2 \to 5_2^3 \to 5_2^4. \tag{30}$$

The orbit starts with the NS5-brane, which is completely geometric and whose background can be consistently described both locally and globally in terms of the metric and the gauge field. Performing T-duality transformations along smeared transverse directions one obtains KK5-monopole 5_2^1 and then the exotic 5_2^2-brane. The background of the 5_2^2-brane cannot be described globally as it has non-trivial monodromy and is glued by T-duality. Going further along the orbit one recovers 5_2^3-branes, whose background is not well defined even locally, and 5_2^4-brane which is object of co-dimension-0. These branes, the corresponding backgrounds and their description in terms of T-duality covariant field theory will be considered in more details in Section 3.1.

Branes of M-theory completely wrapped on d compact direction with radii R_i are in correspondence with massive BPS states of maximal $(11 - d)$-dimensional supergravity, and can be labelled $b^{(c_r...c_1)}$ similarly to the branes of string theory. Tension of $b^{(c_r...c_1)}$-brane, or equivalently mass of the corresponding state, is then given by

$$M\left[b^{(c_r...c_1)}\right] = \frac{R_{i_1}\ldots R_{i_b} R_{j_1}^2 \ldots R_{j_{c_1}}^2 \ldots}{l_{11}^{b+2c_1+3c_2+\ldots+(r+1)c_r+1}}, \tag{31}$$

where l_{11} is the 11-dimensional Planck mass. In these notations the M2-brane is denotes as $2^0 \equiv 2$, the KK6-monopole is denoted as 6^1.

The exotic branes $5^3, 2^6$ and $0^{(1,7)}$ whose winding modes contribute counting for the $E_{8(8)}$ U-duality group, have three, six and 7 special Hopf-fiber-like direction. In addition, the $0^{(1,7)}$ brane has one cubic special circle. Interpretation of cubic circles is more subtle that that of the quadratic ones, some discussion of these for branes of the Type II theories can be found in [20]. Windings modes for these branes correspond to components of the following mixed symmetry tensors

$$
\begin{aligned}
5^3 &: \quad z_{a_1\ldots a_8, b_1 b_2 b_3} \quad \# = 56 \\
2^6 &: \quad z_{a_1\ldots a_8, b_1\ldots b_6} \quad \# = 28 \\
0^{(1,7)} &: \quad z_{a_1\ldots a_8, b_1\ldots b_7, c_1} \quad \# = 8
\end{aligned}
\qquad (32)
$$

where indices to the right after commas must be equal to the those to the left, i.e., for the last line we have $c_1 = b_1 = a_1$ and b's = a's. One immediately notices the reason these branes are the ones which are able to contribute the counting of winding for the $E_{8(8)}$ theory. Indeed, this U-duality group corresponds to $d = 8$ toroidal directions, hence 8 is the maximal number of antisymmetric indices one can have in a winding mode. Say for the 5^3 brane with 3 special directions this provides room for 5 longitudinal directions, while for the $0^{(1,7)}$ one is left with none, hence a 0-brane.

It is important to note that although one includes windings of exotic branes into the counting for the $E_{8(8)}$ case, components of the generalized momentum still do not sum up to 248, the smallest possible irrep being rather 240. This cannot be cured by adding more exotic branes, since no other such brane is able to fit the 8-torus. More importantly, as it will be clear from analysis of the field spectrum of exceptional field theory, the problem is much more than just counting of extra coordinates, a huge part of which would be dropped anyway. Vector fields of the theory must be in the same irrep as the generalized momentum to add up to a covariant derivative, hence one needs extra 8 field to build up an irrep. This is indeed what happens in the $E_{8(8)}$ ExFT: one introduces extra tensor fields and imposes conditions on them similar to the section constraint [21]. For $E_{9(9)}$, which is an affine algebra with infinite-dimensional representations, one has to add infinitely many coordinates and fields. For larger U-duality groups the problem is even more complicated as first one encounters E_{10}, which is an extension of the E_8 by two imaginary roots, and then E_{11}, which must encode timelike direction as well. Some progress in the construction of such infinitely dimensional extended spaces and the corresponding theories has been made in [22–25].

2.3. Exceptional Field Theories

Consider a general exceptional field theory defined on D-dimensional space-time parametrized by coordinates x^μ, which we will call "external" and N-dimensional extended space, which we will call "internal" parametrized by $\mathbb{X}^M$. It is important to note, however, that no compactification is assumed in formulation of the theory. Toroidal reductions of supergravity described in the previous subsection have been used exclusively for counting winding modes. In the framework of exceptional field theories, toroidal backgrounds represent a solution, which preserves the maximal amount of U-duality. In this respect, toroidal backgrounds for ExFT are the same as Minkowski for General Relativity. In what follows, all fields are allowed to depend on the whole set of coordinates $(x^\mu, \mathbb{X}^M)$.

On this space the following field content is defined: external metric $g_{\mu\nu}$, N vector fields $A_\mu{}^M$, generalized metric $\mathcal{M}_{MN}$ and various external tensors $\mathcal{B}_{\mu_1\ldots\mu_r}{}^\alpha$ transforming under an irrep of G the U-duality group. Generalized metric parametrized the coset space G/K, where K is the maximal compact subgroup of G. For supersymmetric theories one in addition defines fermions, transforming under an irrep of K.

Locally geometry of the extended space parametrized by $\mathbb{X}^M$ is represented by the generalized Lie derivative $\mathcal{L}_\Lambda$, which defines generalized diffeomorphism transformations of generalized tensors along a generalized vector Λ^M. Since, the vector $\Lambda^M = \Lambda^M(x, \mathbb{X})$ depends both on the external and internal coordinates, partial derivative ∂_μ of a generalized tensor does not transform covariantly. To fix that one introduces long derivative following the usual approach of the Yang-Mills theory

$$D_\mu = \partial_\mu - \mathcal{L}_{A_\mu}. \tag{33}$$

Hence, the vector fields $A_\mu{}^M$ play the role of gauge connection in the theory. The corresponding field strength is defined in the usual way

$$[D_\mu, D_\nu] = -\mathcal{L}_{F_{\mu\nu}}. \tag{34}$$

One however faces a subtlety here noticing that such defined field strength $F_{\mu\nu}{}^M$ is not a generalized tensor. This situation is familiar from the construction of maximal gauged supergravities, where one has to add p-form gauge field to a p-form field strength to build covariant expressions. Eventually, this leads to tensor hierarchy of the theory [26]. Hence, one defines

$$\mathcal{F}_{\mu\nu}^M = F_{\mu\nu}{}^M + Y_{KL}^{MN}\partial_N B_{\mu\nu}{}^{KL}, \tag{35}$$

using that $\mathcal{L}_{\mathcal{F}_{\mu\nu}} \equiv \mathcal{L}_{F_{\mu\nu}}$ up to the section constraint. In general, one shows that generalized vectors of the form $\Lambda_0^M = Y_{KL}^{MN}\partial_N \chi^{KL}$ for any χ^{KL} do not induce generalized diffeomorphisms when the section constraint is satisfied (for details see [27]). Hence, the 2-forms of the theory came from reduction of the 3-form of 11-dimensional supergravity are utulized in the covariant field strength.

If the 2-forms are dynamical, one needs to construct a covariant field strength for them, which will enter the fully covariant Lagrangian. For that, one considers Bianchi identity for the 2-form field strength, which can be written as

$$3D_{[\mu}\mathcal{F}_{\nu\rho]}{}^M = Y_{KL}^{MN}\partial_N H_{\mu\nu\rho}{}^{KL}. \tag{36}$$

As before, such recovered field strength $H_{\mu\nu\rho}{}^{KL}$ is covariant only when contracted with the Y-tensor, and hence needs additional contributions. One invokes 3-form potential $C_{\mu\nu\rho M,KL}$, for whose field strength one needs a 4-form and so on. This tower of p-forms and their field strengths is called tensor hierarchy of exceptional field theories and drops to that of maximal gauged supergravities upon the Scherk-Schwarz reduction ansatz (see below). In principle, the hierarchy can go up to the top form; however it ends much earlier since at some point one no longer needs to construct yet another field strength since the corresponding gauge field is non-dynamical and does not enter the Lagrangian. Although the E_7 ExFT is more suitable for phenomenological applications, it has short tensor hierarchy and issues with self-duality. Hence, let us consider the construction of the SL(5) ExFT in more detail to illustrate the general idea.

Covariant field strength for the gauge field $A_\mu{}^{mn}$ that is recovered from the commutator of long derivatives and then shifted by a derivative of $B_{\mu\nu m}$, takes the following form [28]

$$\mathcal{F}_{\mu\nu}^{mn} = 2\partial_{[\mu}A_{\nu]}^{mn} - [A_\mu, A_\nu]_C^{mn} + \frac{1}{4}\epsilon^{mnklp}\partial_{kl}B_{\mu\nu p}, \tag{37}$$

where ϵ^{mnklp} is the fully antisymmetric invariant tensor of SL(5). Explicit check shows that under generalized transformations of the fundamental fields $A_\mu{}^{mn}$ and $B_{\mu\nu m}$ such defined field strength transforms covariantly, i.e., $\delta_\Lambda \mathcal{F}_{\mu\nu}{}^{mn} = \mathcal{L}_\Lambda \mathcal{F}_{\mu\nu}{}^{mn}$.

Bianchi identity for the 2-form field strength has non-trivial RHS and which can be written as a full derivative of a tensor $\mathcal{F}_{\mu\nu\rho m}$

$$3\mathcal{D}_{[\mu}\mathcal{F}_{\nu\rho]}{}^{mn} = -\frac{1}{16}\epsilon^{imnkl}\partial_{kl}\mathcal{F}_{\mu\nu\rho i}. \tag{38}$$

This tensor however has a freedom in adding of terms of the type

$$\partial_{mn} \mathcal{C}_{\mu\nu\rho}{}^m. \tag{39}$$

Indeed, in the Bianchi identity such term will give $\epsilon^{imnkl}\partial_{kl}\partial_{ij}\mathcal{C}_{\mu\nu\rho}{}^j \equiv \epsilon^{imnkl}\partial_{[kl}\partial_{ij]}\mathcal{C}_{\mu\nu\rho}{}^j$, which vanishes upon the section constraint of the theory

$$\epsilon^{imnkl}\partial_{mn} \bullet \partial_{kl} \bullet = 0. \tag{40}$$

To match tensor hierarchy of the maximal $D = 7$ gauged supergravity [29] one fixes a coefficient in front of the term and the covariant 3-form field strength reads

$$\mathcal{F}_{\mu\nu\rho m} = 3\mathcal{D}_{[\mu} B_{\nu\rho]m} + \frac{3}{2}\epsilon_{mpqrs}\left(A^{pq}_{[\mu}\partial_\nu A^{rs}_{\rho]} - \frac{1}{3}[A_{[\mu}, A_\nu]_E{}^{pq} A_{\rho]}{}^{rs}\right) - \frac{1}{4}\partial_{mn}\mathcal{C}_{\mu\nu\rho}{}^n. \tag{41}$$

Now one notices by looking at the Figure 2 that there are too little degrees of freedom to compose both the 2-form and the 3-form gauge potentials. On the other hand, according to the prescription of [11] the three form $\mathcal{C}_{\mu\nu\rho}$ should be dualized into a two form to complete the irrep **5** of SL(5). Speaking about covariant theories, one would like to keep information about the 3-form, which is done by taking into account an alternative prescription, which is to dualize the 2-forms $B_{\mu\nu a}$ into 3-forms to complete the irrep $\bar{\mathbf{5}}$ of SL(5). Hence, one ends up with the same degrees of freedom encoded in the fields $B_{\mu\nu m}$ and $\mathcal{C}_{\mu\nu\rho}{}^m$. The duality relation between these potentials can be written in the following covariant form

$$m^{mn}\mathcal{F}^{\mu\nu\rho}{}_n - \frac{1}{4!}\epsilon^{\mu\nu\rho\lambda\sigma\tau\kappa}\mathcal{F}_{\lambda\sigma\tau\kappa}{}^m = 0, \tag{42}$$

where $\epsilon^{\mu\nu\rho\sigma\kappa\lambda\tau}$ is the Levi-Civita tensor in the external 7 directions and m^{mn} is the inverse of the generalized metric m_{mn}.

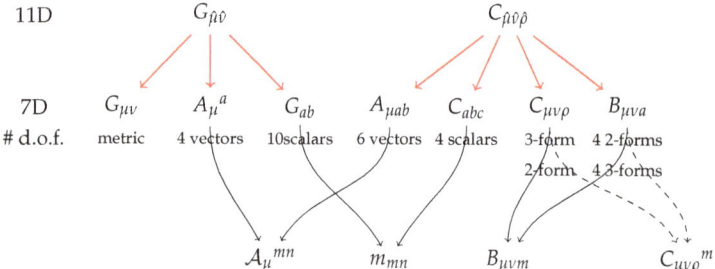

Figure 2. Reduction of 11D fields into seven dimensions, dualization into the lowest possible rank tensor and recollection into SL(5) multiplets. Dashed lines correspond to an alternative combination of the 2- and 3-forms into the 3-form $\mathcal{C}_{\mu\nu\rho}{}^m$ dual to $B_{\mu\nu m}$ in 7 dimensions (see the text). Here the small Latin indices $m, n, k, \ldots$ label the fundamental irrep **5** of SL(5) and small Latin indices from the beginning of the alphabet label the fundamental irrep **4** of GL(4).

Since the 3-form field strength contains the same information as the 4-form field strength, the latter is considered to be non-physical and is not included into the Lagrangian. It appears only in the corresponding Bianchi identity

$$4\mathcal{D}_{[\mu}\mathcal{F}_{\nu\rho\sigma]m} = 6\epsilon_{mpqrs}\mathcal{F}_{[\mu\nu}{}^{pq}\mathcal{F}_{\rho\sigma]}{}^{rs} + \partial_{nm}\mathcal{F}_{\mu\nu\rho\sigma}{}^n \tag{43}$$

and is needed to ensure invariance of the Lagrangian with respect to external diffeomorphisms. Note that the 3-form gauge potential containing in the 4-form field strength has been already used for construction of the 3-form covariant field strength (see Figure 2).

Lagrangian of the theory is schematically the same as the one for the maximal $D = 7$ gauged supergravity

$$\mathcal{L}_{EFT} = \mathcal{L}_{EH}(\hat{R}) + \mathcal{L}_{sc}(\mathcal{D}_\mu m_{kl}) + \mathcal{L}_V(\mathcal{F}_{\mu\nu}{}^{mn}) + \mathcal{L}_T(\mathcal{F}_{\mu\nu\rho\,m}) \\ + \mathcal{L}_{top} - V(m_{kl}, g_{\mu\nu}). \tag{44}$$

Here the kinetic part is given by the Einstein-Hilbert term, scalar kinetic term, and vector and tensor kinetic terms respectively. In addition, one includes the "scalar potential" V, which contains only terms with derivatives ∂_{mn} along extended coordinates, and the topological term $\mathcal{L}_{top}$ which does not contain contractions with the metric. Explicitly the expressions read

$$\begin{aligned}
\mathcal{L}_{EH} &= e e_a^\mu e_\beta^\nu \hat{R}_{\mu\nu}{}^{\alpha\beta}, \\
\mathcal{L}_{sc} &= \frac{1}{12} e g^{\mu\nu} \mathcal{D}_\mu \mathcal{M}_{MN} \mathcal{D}_\nu \mathcal{M}^{MN}, \\
\mathcal{L}_V &= -\frac{1}{4} e \mathcal{M}_{MN} \mathcal{F}_{\mu\nu}{}^M \mathcal{F}^{\mu\nu\,N} = -\frac{1}{8} e m_{mk} m_{nl} \mathcal{F}_{\mu\nu}{}^{mn} \mathcal{F}^{\mu\nu kl}, \\
\mathcal{L}_T &= -\frac{1}{3 \cdot (16)^2} e m^{mn} \mathcal{F}_{\mu\nu\rho\,m} \mathcal{F}^{\mu\nu\rho}{}_n \\
V &= -\frac{1}{13} \mathcal{M}^{MN} \partial_M \mathcal{M}^{KL} \partial_N \mathcal{M}_{KL} + \frac{1}{2} \mathcal{M}^{MN} \partial_M \mathcal{M}^{KL} \partial_L \mathcal{M}_{NK} \\
&\quad - \frac{1}{2}(g^{-1} \partial_M g) \partial_N \mathcal{M}^{MN} - \frac{1}{4} \mathcal{M}^{MN} (g^{-1} \partial_M g)(g^{-1} \partial_N g) - \frac{1}{4} \mathcal{M}^{MN} \partial_M g^{\mu\nu} \partial_N g_{\mu\nu}.
\end{aligned} \tag{45}$$

Here the modified Riemann curvature $\hat{R}_{\mu\nu ab} = R_{\mu\nu ab} + \mathcal{F}_{\mu\nu}{}^M e_a^\rho \partial_M e_{\rho\beta}$ transforms covariantly under both generalized and external diffeomorphisms. The potential V is written in terms of the 10×10 generalized metric $\mathcal{M}_{MN}$ which is related to the 5×5 metric m_{mn} by

$$\mathcal{M}_{mn\,kl} = m_{mk} m_{nl} - m_{ml} m_{nk}, \tag{46}$$

i.e., each capital index M labelling the **10** is represented by an antisymmetric pair of indices $[mn]$ each labelling the fundamental **5** of $SL(5)$. In addition, one halves each contraction of capital indices when turning to pairs to avoid extra counting.

The topological part cannot be written in a covariant form, precisely as in the gauged supergravity case; however its variation can

$$\delta \mathcal{L}_{top} = \frac{1}{16 \cdot 4!} \epsilon^{\mu\nu\rho\lambda\sigma\tau\kappa} \left[\mathcal{F}_{\mu\nu\rho\lambda}{}^m \partial_{mn} \Delta C_{\sigma\tau\kappa}{}^n + 6 \mathcal{F}_{\mu\nu}{}^{mn} \mathcal{F}_{\rho\lambda\sigma m} \Delta B_{\tau\kappa n} - 2 \mathcal{F}_{\mu\nu\rho m} \mathcal{F}_{\lambda\sigma\tau n} \delta A_\kappa{}^{mn} \right]. \tag{47}$$

Transformations Δ of the fields denote external diffeomorphisms parametrized by a vector ξ^μ, generalized diffeomorphisms parametrized by a generalized vector Λ^{mn} and gauge transformations of the fields $A_\mu{}^{mn}, B_{\mu\nu m}, C_{\mu\nu\rho}{}^m$:

$$\begin{aligned}
\delta e_\mu^a &= \xi^\nu \mathcal{D}_\nu e_\mu^a + \mathcal{D}_\mu \xi^\nu e_\nu^a, \\
\delta \mathcal{M}_{MN} &= \xi^\mu \mathcal{D}_\mu \mathcal{M}_{MN}, \\
\delta A_\mu{}^M &= \xi^\nu \mathcal{F}_{\nu\mu}{}^M + \mathcal{M}^{MN} g_{\mu\nu} \partial_N \xi^\nu + \mathcal{D}_\mu \Lambda^{mn} + \frac{1}{16} \epsilon^{imnkl} \partial_{kl} \Xi_{\mu i}, \\
\Delta B_{\mu\nu i} &= \xi^\rho \mathcal{F}_{\rho\mu\nu i} + 2 \mathcal{D}_{[\mu} \Xi_{\nu]i} - 2 \epsilon_{imnpq} \Lambda^{mn} \mathcal{F}_{\mu\nu}{}^{pq} - \partial_{mi} \Psi_{\mu\nu}{}^m, \\
\Delta C_{\mu\nu\rho}{}^m &= -\frac{1}{3!} e \epsilon_{\mu\nu\rho\sigma\kappa\lambda\tau} \xi^\sigma m^{mn} \mathcal{F}^{\kappa\lambda\tau}{}_n + 3 \mathcal{D}_{[\mu} \Psi_{\nu\rho]}{}^m + 3 \mathcal{F}_{[\mu\nu}{}^{mn} \Xi_{\rho]n} + \Lambda^{mn} \mathcal{F}_{\mu\nu\rho n}.
\end{aligned} \tag{48}$$

The amazing feature of exceptional field theories observed already in [30] is that the form of the Lagrangian is completely fixed by demanding invariance with respect to the above transformations.

In contrast, in gauged supergravities one has to impose supersymmetry to fix all coefficients, although the general structure of the Lagrangian is completely the same. The non-trivial check is that supersymmetry works well for the Lagrangian of exceptional field theory fixed in such a way. This has been checked explicitly for E_7 and E_6 in [31,32], and is expected to work for all other cases as well.

2.4. The Section Constraint

Exceptional field theories are formulated on a space-time parametrized by d external (conventional) coordinates x^μ and D "internal" extended coordinates $\mathbb{X}^M$. Since no reduction is assumed, these are theories on a $d + D$-dimensional space-time, which always has dimension greater than 11. To reduce the amount of degrees of freedom, and to ensure consistency of algebra of generalized diffeomorphism one has to impose condition, the section constraint, which is

$$Y^{MN}_{KL} \partial_M \bullet \partial_N \bullet, \tag{49}$$

where $\bullet$ stands for any field or combination of fields. Y-tensors for different U-duality groups are listed in Table 1.

This condition restricts dependence of fields of the theory on the extended coordinates and can be solved in various ways. Consider as an example the E_7 theory for which the condition is

$$\begin{aligned} \Omega^{MN} \partial_M \bullet \partial_N &= 0, \\ t_\alpha{}^{MN} \partial_M \bullet \partial_N &= 0 \end{aligned} \tag{50}$$

where Ω^{MN} is the invariant symplectic form of $E_7 < Sp(56)$ and t_α are generators of the group. In [33] it has been shown that this condition has precisely two algebraic types of solutions: corresponding to embeddings of the D = 11 maximal supergravity and of Type IIB supergravity. All other follow from this two. In addition, there is a special way to relax this constraint by imposing the so-called Scherk-Schwarz ansatz.

Consider first the algebraic solution of the section constraint corresponding to embedding of the 11D maximal supergravity. In this case, one decomposes the fundamental **56** of E_7 under the action of the $GL(7)$ subgroup as

$$\mathbf{56} \to \mathbf{7}_{+3} + \mathbf{21}'_{+1} + \mathbf{21}_{-1} + \mathbf{7}'_{-3}, \tag{51}$$

where subscripts denote the GL(1) weight. For the coordinates $\mathbb{X}^M$ this implies

$$\mathbb{X}^M = (x^m, y_{mn}, \tilde{y}^{mn}, \tilde{x}_m). \tag{52}$$

Here and further in this subsection

$$\begin{aligned} M, N, K, \ldots &= 1, \ldots, 56, & &\text{label the } \mathbf{56} \text{ of } E_7, \\ m, n, k \ldots &= 1, \ldots, 7, & &\text{label the } \mathbf{7} \text{ of } SL(7), \\ a, b, c, d, \ldots &= 1, \ldots, 6, & &\text{label the } \mathbf{6} \text{ of } SL(6), \\ \alpha, \beta, \ldots &= 1, 2, & &\text{label the } \mathbf{2} \text{ of } SL(2). \end{aligned} \tag{53}$$

The section constraint then can be satisfied by imposing

$$\frac{\partial}{\partial y_{mn}} = 0, \quad \frac{\partial}{\partial \tilde{y}^{mn}} = 0, \quad \frac{\partial}{\partial \tilde{x}_m} = 0, \tag{54}$$

hence fields depend only on 7 coordinates x^m and 4 coordinates x^μ, in total 11. Field content of the E_7 theory decomposed under the GL(7) subgroup apparently fits the field content of the 11 dimensional supergravity in the 7 + 4 split form, as the former was constructed from the latter initially. Type IIA supergravity is obtain from this theory by further $\mathbb{S}^1$ reduction in the usual way.

The alternative solution of the section constraint is associated with decomposition under the embedding $E_7 \hookleftarrow GL(6) \times SL(2)$, which implies for the fundamental irrep

$$56 \to (6,1)_{+2} + (6',2)_{+1} + (20,1)_0 + (6,2)_{-1} + (6',1)_{-2}. \tag{55}$$

For components of the extended coordinates this reads

$$\mathbb{X}^M = (x^a, y_{a\alpha}, y_{abc}, \tilde{y}^{a\alpha}, \tilde{x}_a), \tag{56}$$

and the field content will be decomposed accordingly, The SL(2) subgroup transforming subsets of dual coordinates correspond to S-duality symmetry of Type IIB supergravity. The section constraint implies

$$\frac{\partial}{\partial y_{a\alpha}} = 0, \quad \frac{\partial}{\partial y_{abc}} = 0, \quad \frac{\partial}{\partial \tilde{y}^{a\alpha}} = 0, \quad \frac{\partial}{\partial \tilde{x}_a} = 0, \tag{57}$$

and the fields depend on 6 coordinates x^a and 4 coordinates x^μ, in total 10. After reduction, the SL(2) S-duality symmetry can be kept manifestly at the cost of the full 10-dimensional Lorenz symmetry. To restore the latter, one has to break the former.

A special solution to the section constraint trivially embedded into the above two classes is $\partial_M = 0$, i.e., all fields do not depend on extended coordinates. This corresponds simply to d-dimensional maximal ungauged supergravity, i.e., a reduction of the 11-dimensional supergravity on a torus $\mathbb{T}^{11-d}$. These are known to allow deformations, gaugings, producing a class of theories corresponding to various reductions of the initial 11-dimensional theory [26]. Exceptional field theories are able to reproduce such gauged supergravities as well under a special ansatz, which relaxes the differential section condition to an algebraic constraint. These will be considered in more detail in Section 3.1.

2.5. Double Field Theory

Double field theory, which is a T-duality covariant formulation of supergravity, also fits the above scheme. The duality group is O(n,n) with $n + d = 10$, the Y-tensor then becomes $Y^{MN}_{KL} = \eta^{MN}\eta_{KL}$, where $M, N, K, L = 1, \ldots, 2n$ for any n. Since algebraic structure of the theory does not depend on the number of "internal" dimensions, it is natural to extend all 10 coordinates of Type II supergravity. The generalized metric then

$$M_{MN} \in \frac{O(10,10)}{O(1,9) \times O(9,1)}, \tag{58}$$

and the rest of the construction is repeated identically.

Important remark here concerns the section constraint, which is

$$\eta^{MN} \partial_M \bullet \partial_N \bullet = \partial_m \bullet \tilde{\partial}^m \bullet = 0, \tag{59}$$

where we introduce notation $\mathbb{X}^M = (x^m, \tilde{x}_m)$ for extended coordinates. The above implies that for any pair of a normal coordinate x^* and the corresponding dual coordinate $\tilde{x}_*$ fields are allowed do depend on one or another, but not on both. The choices are related by a T-duality transformation precisely as in the Duff's procedure

$$T_* : x^* \longleftrightarrow \tilde{x}_*. \tag{60}$$

Hence, such defined T-duality means that one translates all upper indices $*$ into lower indices $*$, and the coordinates $\tilde{x}_*$ becomes normal geometric coordinate in the new T-duality frame. Note however a different possibility, where one simply replaces dependence of background fields on x^* by $\tilde{x}_*$, still counting the latter as a non-geometric dual coordinate. This will turn a solution of supergravity equations of motion into a proper string background, which however (i) does not solve e.o.m.s,

(ii) is non-geometric. This way DFT and ExFT allow addressing exotic branes and the corresponding non-geometric backgrounds. See further Section 3.2 for more details.

3. Applications

3.1. Non-Geometric Compactifications

Choosing a solution of the section constraint of exceptional field theory can be understood as a dimensional reduction as it drops dependence of fields of the theory on a subset of coordinates. E.g., the trivial solution $\partial_M \bullet = 0$ when all fields depend only on D external coordinates corresponds to reduction of the 11-dimensional supergravity on a $11 - D$-torus. Other toroidal reductions can as well be obtained in such a way by choosing a smaller subset of the extended coordinates to be dropped. However, exceptional field theory is able to incorporate more general reductions, in fact all parametrized in terms of embedding tensor.

Taking the general idea of twisted dimensional reduction of Scherk and Schwarz [34] one introduces the following generalized Scherk-Schwarz ansatz

$$V^M(x, \mathbb{X}) = U^M{}_A(\mathbb{X}) V^A(x), \tag{61}$$

where $U^M{}_A(\mathbb{X})$ is usually referred to as the twist matrix. The ansatz tells that all information about dependence on the extended coordinates $\mathbb{X}^M$ is contained in the twist matrix. Inserting this into the definition of generalised Lie derivative one obtains

$$\mathcal{L}_\Lambda V^M = U^M{}_A(\mathbb{X}) X_{BC}{}^A \Lambda^A(x) V^B(x), \tag{62}$$

which can be thought of as a definition of the tensor $X_{AB}{}^C$

$$X_{AB}{}^C \equiv 2 U_M{}^C U_{[A}{}^N \partial_N U_{B]}{}^M + Y_{EB}^{CD} U_M{}^E U_D{}^N \partial_N U_A{}^M, \tag{63}$$

where $U_M{}^C U_C{}^N = \delta_M{}^N$. In principle, this tensor depends on $\mathbb{X}^M$ as it is constructed out of twist matrices and their derivatives; however, for now we will assume $X_{AB}{}^C = $ const. Certainly this implies restrictions on the twist matrices which will be discussed in a moment.

Upon the generalised Scherk-Schwarz ansatz the closure constraint (23) and the condition for $X_{AB}{}^C$ to transform covariantly boil down to the following simple algebraic constraint

$$[X_A, X_B] = -X_{AB}{}^C X_C. \tag{64}$$

Here we understand $X_{AB}{}^C$ as a matrix labelled by A. The above looks as a commutation relation for an algebra with generators X_A; however, since $X_{AB}{}^C$ is not necessarily antisymmetric, such simple interpretation does not work. Instead, one recognizes here the structure of gauged supergravities, with $X_{AB}{}^C$ being the embedding tensor. Hence, the main line of the subsection is that Scherk-Schwarz reduction of exceptional field theories replaces the differential section constraint by the algebraic condition (64), which has precisely the same form as the quadratic constraint of gauged supergravity. Moreover, performing the reduction at the level of the action one reproduces precisely action of the corresponding D-dimensional gauged supergravity. This has been shown first for DFT in [35] and then for exceptional field theories SL(5), SO(5,5), E_6, E_7 and the enhanced O(d,d) exceptional field theory in [36–39]. Let us now briefly describe structure of gauged supergravities and their relation to generalized Scherk-Shwarz reductions.

Gauged supergravities most straightforwardly can be described in the so-called embedding tensor formalism first developed in the context of three dimensional theories [40,41] and then constructed for other maximal supergravities (for review see [26,42,43]). When dimensionally reduced on a d-torus 11-dimensional supergravity produces maximally supersymmetric theory in $D = 11 - d$ dimensions, which contains n_V vector fields $\mathcal{A}_\mu{}^A$, with $A = 1, \ldots, n_V$. The vector fields descent from the metric and the 3-form C-field in 11 dimensions precisely in the same way as the generalized vector fields of exceptional field theory (see Figures 1 and 2). For toroidal reductions the resulting theory is abelian with the gauge group $U(1)^{n_V}$ and gauge transformations

$$\delta_\Lambda A_\mu{}^A = \partial_\mu \Lambda^A. \tag{65}$$

In addition, the lower dimensional theory has U-duality symmetry group G and the vector multiplet transforms under an irrep $\mathcal{R}_V$ of dimension $\dim \mathcal{R}_V = n_V$. Reductions on more complicated manifolds endowed with torsion and/or curvature, and reductions in the presence of fluxes of gauge fields result in theories with less symmetry, and with vector multiplets belonging to adjoint representation of a non-abelian gauge group. Such reductions correspond to the diagonal arrow on Figure 3, while toroidal reductions correspond to the vertical arrow. Hence, it is natural to deform the abelian maximal theory introducing non-abelian interactions between the n_V vector fields, preserving supersymmetry and local symmetries of the theory. A self-consistent algorithm for such a procedure is based on the notion of embedding tensor Θ, which defines embedding of the desired local gauge group G' into the full global U-duality group G

$$\Theta : G \to G'. \tag{66}$$

One writes deformation of the gauge transformation rule for vector multiplets as follows

$$\delta_\Lambda A_\mu{}^A = \partial_\mu \Lambda^A - g X_{BC}{}^A \Lambda^B A_\mu{}^C, \tag{67}$$

where g is the deformation parameter. The "structure constants" $X_{AB}{}^C$ can be written in terms of the embedding tensor as

$$X_{AB}{}^C = Z_{AB}{}^C + \hat{X}_{AB}{}^X = \Theta_A{}^\alpha t_{\alpha B}{}^C, \tag{68}$$

where $\{t_\alpha\} = $ bas $\mathfrak{g}$ is basis of generators of the global U-duality group G, and Z and $\hat{X}$ are symmetric and antisymmetric in $\{AB\}$ respectively. Hence, the embedding tensor indeed selects a subset of these generators to construct the constants $X_{AB}{}^C$, which define the non-abelian gauge transformation.

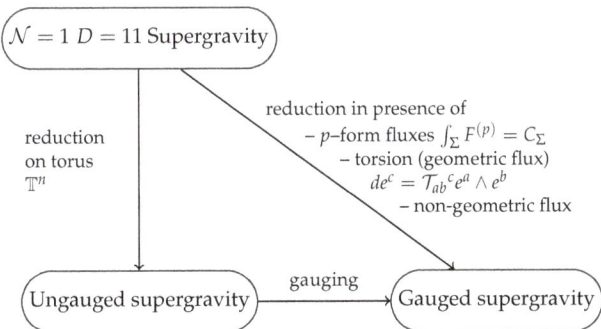

Figure 3. Relations between toroidal reductions of $\mathcal{N} = 1\, D = 11$ supergravity, gaugings and more complicated dimensional reductions involving geometric and non-geometric fluxes.

Certainly, such a deformation of the ungauged D-dimensional theory cannot be done in an arbitrary way and has rather to satisfy certain constraints. The first one is the linear constraint and comes from the condition that the resulting deformed equations of motion are still invariant under the full set of supersymmetries, i.e., the deformed gauged supergravity is a maximally supersymmetric theory. The linear constraint projects certain representations of G in the decomposition of the embedding tensor $\Theta_A{}^\alpha \in \mathcal{R}_{n_V} \otimes \text{adj}\,\mathfrak{g}$, which can be schematically written as

$$\mathbb{P}\Theta = 0. \tag{69}$$

As an example, one may consider the $D = 4$ theory constructed in [13,44], for which $G = \mathbb{R}^+ \times E_7$, $\mathcal{R}_{n_v} = \mathbf{56}$, $\text{adj}\,\mathfrak{g} = \mathbf{133}$ and

$$\Theta_A{}^\alpha \in \mathbf{56} \otimes \mathbf{133} = \mathbf{56} + \mathbf{912} + \mathbf{6480}. \tag{70}$$

The linear constraint tells that for the theory to be maximally supersymmetric the embedding tensor cannot have the $\mathbf{6480}$ part

$$\mathbb{P}_{\mathbf{6480}}\Theta = 0. \tag{71}$$

Please note that the duality group to be gauged contains a $\mathbb{R}^+$ part, which correspond to global rescalings of the external metric. After the gauging procedure the corresponding generators correspond to the so-called trombone gauging. Certainly, theories with non-zero trombone part of the embedding tensor cannot be written in terms of an action, as rescalings are symmetries of equations of motion, rather than the action [44].

In addition to the linear constraint the embedding tensor must respect the so-called quadratic constraint, which follows from the condition of covariancy of the tensor under global transformations. These are nothing but (64)

$$[X_A, X_B] = -X_{AB}{}^C X_C. \tag{72}$$

Note that the LHS does not have the part symmetric in $\{A, B\}$, while the tensor $X_{AB}{}^C$ has no symmetry in the lower indices in general. Hence, the above implies constraint on the symmetric part $Z_{AB}{}^C X_{CD}{}^E = 0$. This condition is essential for the Jacobi identity for gauge transformations. Indeed, one discovers that the Jacobiator of gauge transformations has the conventional form

$$[\delta_{\Lambda_1}, \delta_{\Lambda_2}, \delta_{\Lambda_3}]V^A = -3\hat{X}_{[BC}{}^G \hat{X}_{D]G}{}^E X_{EF}{}^A \Lambda_1 \Lambda_2 \Lambda_3 V^F. \tag{73}$$

However, the commutation relations (64) restricted for the antisymmetric part only do imply the RHS of the above expression to vanish, which is usually the case for a conventional Lie algebraic structure constants. Instead one again finds the symmetric part

$$3\hat{X}_{[BC}{}^G \hat{X}_{D]G}{}^E = Z_{G[B}{}^E X_{CD]}{}^G. \tag{74}$$

hence, the Jacobi identity is satisfied upon $Z_{AB}{}^C X_C = 0$ similarly to the closure constraint.

Such rich algebraic structure of the deformations means that one is not simply dealing with a single gauge group, but rather with a set of gauge groups inside the full duality group G, defined by choosing the embedding tensor. Gauged supergravities for different dimensions have been constructed in the series of papers [13,29,44–47] which contain detailed analysis of the constraint briefly described above, construction of the corresponding effective action including the fermionic sector and analysis of some examples.

Solving the linear and quadratic constraints for the embedding tensor one ends up with a theory which can be in principle obtained as a dimensional reduction of the 11-dimensional supergravity on some manifold. For example, a particular choice of gaugings for the $D = 4$ theory correspond to embedding $\Theta : G \to SO(8)$, which gives reduction on a $\mathbb{S}^7$ sphere considered in [48,49]. In this case, components of the embedding tensor are equal to the integral value of the flux of the 7-form field strength on the sphere.

Such gaugings, which can be interpreted in terms of dimensional reductions are called geometric and are said to have higher dimensional origin. However, the full set of solutions to the quadratic and linear constraint contains gaugings which do not have apparent higher dimensional origin, moreover one can show that some cannot have any [50]. Although these still define consistent theories in lower dimensions and give masses and couplings for lower dimensional fields, one cannot come up with a dimensional reduction scheme starting from the 11D supergravity and ending up with these theories.

At this point, one turns back to the beginning of this subsection, where Scherk-Schwarz reduction of exceptional field theories have been shown to produce algebraic structures of the same type as those of gauged supergravities. In principle, since any gauging is given as an expression written in terms of twist matrices and their derivatives, as in (63), one may hope that it is possible to solve these equations and present twist matrices explicitly for each type of gauging. This has been shown to be true for $D = 7$ half-maximal gaugings and $D = 8$ maximal gaugings in [50], where full classification of the corresponding twist matrices was obtained. The most outstanding output of such analysis is that the so-called non-geometric gaugings, i.e., those which do not have higher standard dimensional origin, can be constructed out of generalized twist matrices. Moreover, the set of non-geometric gaugings itself is divided into normal and genuine non-geometric subsets. The former is defined to belong to a U-duality orbit of a geometric gauging, and hence the theory can be rotated to a frame, where no non-geometry is present. On the other hand, the latter do not belong to such an orbit and hence the theory is always non-geometric. At least for some examples, genuine non-geometric gaugings were shown to descent from twist matrices which *break the section constraint*.

At this point it is suggestive to return to the constraint (64) and recall that for that to satisfy one does not need to impose any conditions on dependence of twist matrices on the extended coordinates. Hence, in principle it is possible to break the section constraint, while satisfying the quadratic constraint that has been shown explicitly in [50]. Certainly, such twist matrices cannot be understood as defining a reduction of the conventional supergravity. Whether such configurations can be understood as proper string backgrounds is not clear, as all duality-covariant setups so far imply the section condition. However, genuine non-geometric gaugings look very promising for cosmological model building as these provide extra parameters in scalar potential of the lower dimensional theory, which may help to stabilize its moduli. Indeed, if one considers a setup with geometric gaugings which fails to stabilize certain subset of scalar moduli, any gaugings belonging to duality orbits of the initial ones would also fail to do so, since the physics must be duality invariant. In contrast, genuine non-geometric gaugings do not suffer from such constraints and hence can enhance the realm of possible models. This approach has been taken to analyse cosmological implications of exceptional field theories in application to some toy-model examples in [51–55].

3.2. Exotic Brane Backgrounds

Geometric gaugings of lower dimensional supergravities are equal to integral values of fluxes on the internal manifold, and hence, naturally acquire interpretation in terms of branes. Wrapped around cycles of the internal space, these source fluxes of gauge p-form fields, which then are subject of the usual Dirac quantisation condition (see e.g., [56] for review). Similarly, non-geometric fluxes can be interpreted as integrated values of field strengths of mixed symmetry potentials sourced by exotic branes. Algebraic-wise one finds good correspondence between mixed allowed symmetry potentials of lower dimensional supergravity (equivalently, exotic branes) and gaugings [57–59]. However, the conventional supergravity is known to be unable to properly describe backgrounds of such exotic branes and to provide a technique for calculating the corresponding non-geometric fluxes.

The well-known example comes from the T-duality orbit of the NS5-brane, starting with the H-flux [18,19]. This is known to contain non-geometric Q- and R-fluxes via the relation

$$H_{abc} \xleftrightarrow{T_a} f_{ab}{}^c \xleftrightarrow{T_b} Q_a{}^{bc} \xleftrightarrow{T_c} R^{abc}, \qquad (75)$$

where T_* denotes T-duality transformation in the corresponding isometry direction. Since in the conventional supergravity T-duality transformation is defined only along a Killing direction and T-duality along the world-volume direction of the NS5-brane do not change the background, one has to smear on the four transverse directions to perform the transformation. The full NS5-brane background is characterized by the harmonic function $H = H(x^1, x^2, x^3, x^4)$ in four Euclidean dimensions

$$\triangle H(x^1, x^2, x^3, x^4) = h\, \delta^{(4)}(x^1, x^2, x^3, x^4). \tag{76}$$

To have flat asymptotics at spatial infinity as $r \to \infty$ one defines the harmonic function as

$$H = 1 + \frac{h}{r^2}, \tag{77}$$

where $r^2 = (x^1)^2 + (x^2)^2 + (x^3)^2 + (x^4)^2$. To smear along say the direction $z := x^4$ one considers harmonic function $H = H(x^1, x^2, x^3)$ which has apparent isometry along z and solves the equations of motion of supergravity with the same amount of flux. This procedure gives the following background of the smeared NS5-brane which we refer to as the H-monopole

$$\begin{aligned} ds^2 &= ds^2_{056789} + H ds^2_{1234}, \\ B &= A \wedge dz, \\ e^{-2(\varphi - \varphi_0)} &= H^{-1}. \end{aligned} \tag{78}$$

where $H = 1 + h/r$ and the 1-form $A = A_i dx^i$ is defined to be of a magnetic configuration

$$2\partial_{[i} A_{j]} = \epsilon_{ijk} \partial_k H, \tag{79}$$

hence the name "H-monopole". Performing T-duality along the compact direction z by simply applying the Buscher rules (11) one arrives at the background of KK-monopole

$$\begin{aligned} ds^2 &= ds^2_{056789} + H ds^2_{123} + H^{-1}(dx^4 + A)^2, \\ B &= 0. \end{aligned} \tag{80}$$

This background has vanishing B-field and the magnetic monopole configuration is given by the metric components $g_{zi} = A_i$. In three transverse directions $\{x^1, x^2, x^3\}$ such background behaves as the gravitational magnetic monopole.

Already at this step the backgrounds start to get tricky. Indeed, the Gross-Perry monopole solution, which describes the $\{t, x^1, x^2, x^3, z\}$ part of the KK5 background, is topologically a Hopf fibration with the special cycle given by the z direction [60]. Moving around this cycle one observes periodicity in $4\pi h$ and glues the background by a diffeomorphisms transformation. This non-trivial topological structure is reflected by non-vanishing geometric flux $f_{ij}{}^z$, defined as structure constants of the algebra of local vielbeins

$$[e_a, e_b] = f_{ab}{}^c e_c, \tag{81}$$

where $a, b = 1, 2, 3, z$ and the vielbein is defined in the usual lower-triangular gauge as

$$e_i{}^a = H^{-\frac{1}{2}} \begin{bmatrix} H & 0 & 0 & 0 \\ 0 & H & 0 & 0 \\ 0 & 0 & H & 0 \\ A_1 & A_2 & A_3 & 1 \end{bmatrix}. \tag{82}$$

To T-dualize further, one has to smear one more direction, say x^3, which brings us to the set of co-dimension-2 solutions. These are not well defined field configurations as one cannot satisfy proper conditions at space infinity asymptotics. Formally, the harmonic function in two dimensions is $H(\rho) = h_0 + h \log \rho/\rho_0$, where $\rho^2 = (x^1)^2 + (x^2)^2$, and diverges both at the core and at the infinity. For that renormalization procedure introduces a dimensionful parameters ρ_0, h_0, which somehow run when one approaches the singularity points [17]. In what follows, we will drop these parameters for simplicity and always assume the harmonic function of the form

$$H = 1 + \tilde{h} \log \rho. \tag{83}$$

Performing T-duality transformation of the KK5-background smeared in such a way one arrives at the following background

$$ds^2 = H(d\rho^2 + \rho^2 d\theta^2) + \frac{H}{H^2 + \tilde{h}^2\theta^2} ds_{34}^2 + ds_{056789}^2,$$

$$B^{(2)} = \frac{\tilde{h}\theta}{H^2 + \tilde{h}^2\theta^2} dx^3 \wedge dx^4, \tag{84}$$

$$e^{-2(\varphi - \varphi_0)} = \frac{H}{H^2 + \tilde{h}^2\theta^2},$$

where $\{\rho, \theta\}$ are the polar coordinates in the transverse plane $\{x^1, x^2\}$. One immediately notices the non-trivial monodromy: when going around the brane $\theta \to \theta + 2\pi$ the background is glued by a T-duality transformation in the directions $\{x^3, x^4\}$. Such non-geometric configurations are believed to be sourced by exotic branes and are properly described on the language of T-folds [15,16,61]. In terms of the classification (28) the background (84) is sourced by the 5_2^2-brane. Now it is easy to see that the two "quadratic" directions in the mass formula correspond to the two special directions of the brane.

The monodromy gluing the torus $\{x^3, x^4\}$ can be conveniently represented as a linear transformation of the corresponding generalized metric

$$\mathcal{H}(\theta' = \theta + 2\pi) = \mathcal{O}^{tr} \mathcal{H}(\theta) \mathcal{O}, \tag{85}$$

where the matrix $\mathcal{O}$ encodes the non-geometric β-transform

$$\mathcal{O} = \begin{bmatrix} 1_2 & 0 \\ \beta(\theta') & 1_2 \end{bmatrix} \tag{86}$$

with $\beta(\theta) = \tilde{h}\theta \, \partial_3 \wedge \partial_4$. This suggests that it is more natural to use the β-frame of DFT [62] where the generalized metric parametrized by the space-time metric and a bi-vector field β^{ab} instead of the 2-form Kalb-Ramond field. In these variables the background becomes

$$ds^2 = H(d\rho^2 + \rho^2 d\theta^2) + H^{-1} ds_{34}^2 + ds_{056789}^2,$$

$$\beta = \beta^{34} \frac{\partial}{\partial x^3} \wedge \frac{\partial}{\partial x^4}. \tag{87}$$

This form naturally reflects the fact that the 5_2^2-brane electrically interacts with the bi-vector field as was shown explicitly in [20,63,64]. Indeed, the bi-vector β^{ab} is actually an 8-form taking values in two-vectors, which can be written in 10-dimensions as

$$\beta_{(8,2)} = \beta_{\mu_1 \dots \mu_8}{}^{v_1 v_2} dx^{\mu_1} \wedge \cdots \wedge dx^{\mu_8} \partial_{v_1} \wedge \partial_{v_2}, \tag{88}$$

with additional condition that all components with any of the upper indices repeating any of the lower indices vanish [58]. Hence, the corresponding Wess-Zumino term for the 5_2^2-brane has the following structure

$$S_{WZ}^{5_2^2} = h \int d^6 \xi \, \beta_{\alpha_1 \dots \alpha_6 12}{}^{34} d\xi^{\alpha_1} \wedge \cdots \wedge d\xi^{\alpha_6}, \tag{89}$$

where ξ^α are the world-volume coordinates.

Turning to the β-frame allows defining the corresponding flux explicitly as $Q_a{}^{bc} = \partial_a \beta^{bc}$, which in 10 dimensions is represented by a (9,2) mixed symmetry tensor. Smearing one more direction x^2 one obtains linearly growing harmonic function, and T-dualising along x^2 arrives at the s-called R-monopole, the 5_2^3-brane. This is a co-dimension-1 solution, a domain wall, which is even more peculiar. Non-vanishing R-flux $R^{abc} = 3\beta^{d[a}\partial_d \beta^{bc]}$ of the background reflects non-associativity of the closed string coordinates on such background [65,66].

One concludes that exotic brane backgrounds in the approach of the conventional supergravity are solutions fo equations of motion of co-dimension ≤ 2, i.e., domain strings, walls and space-time filling objects. Co-dimension-2 solutions are characterized by non-trivial monodromy around the core of the object, which glues the background by a T-duality transformation. In the T-duality covariant approach of DFT one expects to describe a whole duality orbit as a single field configuration. This is indeed possible and has been shown explicitly for the orbit of the NS5-brane in [67,68] and for backgrounds of exotic branes of M-theory in [69–71] (for review see [72]).

In [67,68] it has been shown that all branes, both geometric and exotic, of the T-duality orbit starting with the NS5-brane are just different faces of a single object called DFT monopole. This background is a solution of DFT equations of motion, characterized by the generalized metric written in the form of a quasi-interval $ds^2 = \mathcal{H}_{MN} d\mathbb{X}^M d\mathbb{X}^N$ (note that this is not an invariant expression and cannot be understood as a definition for distance in the doubled space) and the invariant dilaton d

$$\begin{aligned} ds^2_{DFT} = &\, H(1 + H^{-2} A^2) dz^2 + H^{-1} d\tilde{z}^2 + 2H^{-1} A_i (dy^i d\tilde{z} - \delta^{ij} d\tilde{y}_j dz) \\ &+ H(\delta_{ij} + H^{-2} A_i A_j) dy^i dy^j + H^{-1} \delta^{ij} d\tilde{y}_i d\tilde{y}_j \\ &+ \eta_{rs} dx^r dx^s + \eta^{rs} d\tilde{x}_r d\tilde{x}_s, \end{aligned} \quad (90)$$

$$e^{-2d} = H e^{-2\varphi_0}.$$

Here the harmonic function and the vector A_i are that of the H-monopole

$$H(y) = 1 + \frac{h}{\sqrt{\delta_{ij} y^i y^j}}, \quad (91)$$

$$2\partial_{[i} A_{j]} = \epsilon_{ijk} \partial_k H.$$

Of crucial importance here is the understanding of the coordinate dependence of the fields. First, one notes that the section condition is satisfied since the fields depend only on (y^1, y^2, y^3) and do not depend on their duals $(\tilde{y}_1, \tilde{y}_2, \tilde{y}_3)$. Second, the form of the solution as written above does not tell us which coordinates are geometric, i.e., used for measuring physical distances, and which are not. This is additional information which is given upon fixing position of the DFT monopole in the doubled space. Since nothing changes when replacing x^r by $\tilde{x}_r$, equivalently, when T-dualizing along the world-volume directions, one has five possible choices for geometric coordinates x^μ listed in Table 3. Consider the Q-monopole, the 5_2^2 brane, for which the geometric set of transverse coordinates is $(\tilde{z}, \tilde{y}_1, y^2, y^3)$. This implies that the harmonic function, which is always of the form (91), depends on one dual coordinate that is $\tilde{x}_1 = y^1$ in this case. Hence, one has properly defined harmonic function with nice asymptotic behaviour, which however is allowed to depend on dual coordinates. Please note that instead of the H-monopole background one may start with the full NS5-brane background and the harmonic function

$$H(z, y) = 1 + \frac{h}{z^2 + \delta_{ij} y^i y^j}, \quad (92)$$

which would imply that a 5_2^r-brane background is characterized by fields which depend on r dual coordinates. This number if also equal to the number of special circles.

The orbit starting at the NS5-brane was also considered in [67], where the background of the localised Kaluza-Klein monopole has been recovered from the DFT monopole. Such a background has been known before these results and was recovered from the usual Kaluza-Klein monopole background by taking into account backreaction from worldsheet instantons in the two-dimensional linear sigma model [73–76]. Similarly, worldsheet instanton corrections have been shown to localize the background of the 5_2^2 co-dimension-2 brane in dual space [77]. In Figure 4 relations between monopoles and their localized versions are shown schematically.

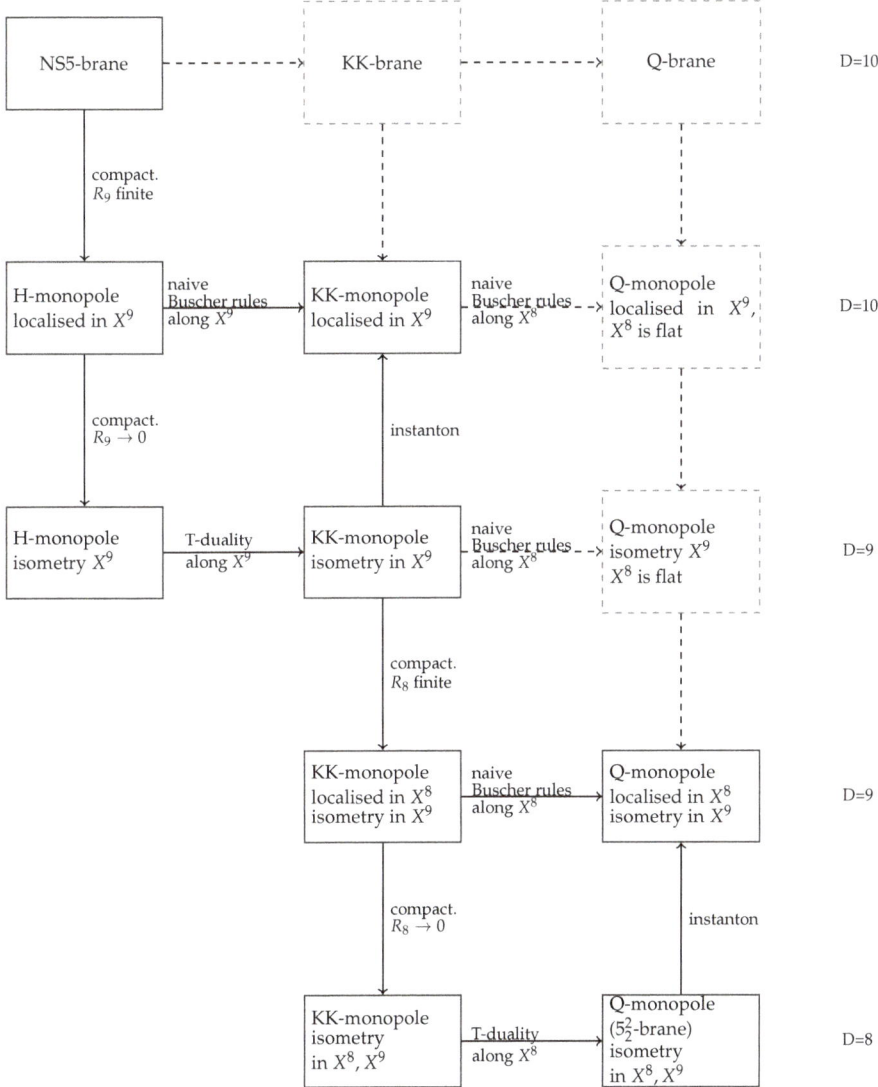

Figure 4. Systematics of backgrounds with H, geometric f and Q fluxes and their relations. Note that in [77] Q-monopole localised in both X^8 and X^9 has been constructed with both these directions being compact.

Table 3. Possible choices for orientation of DFT monopole in the doubled space.

Coordinates	Brane	State
$x^\mu = (z, y^1, y^2, y^3, x^r)$	NS5-brane (H-monopole)	5_2^0
$x^\mu = (\tilde{z}, y_1, y^2, y^3, x^r)$	KK-monopole	5_2^1
$x^\mu = (\tilde{z}, \tilde{y}_1, y^2, y^3, x^r)$	Q-monopole	5_2^2
$x^\mu = (\tilde{z}, \tilde{y}_1, \tilde{y}_2, y^3, x^r)$	R-monopole	5_2^3
$x^\mu = (\tilde{z}, \tilde{y}_1, \tilde{y}_2, \tilde{y}_3, x^r)$	R'-monopole	5_2^4

3.3. Deformations of Supergravity Backgrounds

Backgrounds of DFT depending on a dual coordinate and the possibility to parametrize the generalized metric in terms of exotic degrees of freedom, such as the bi-vector $\beta^{\mu\nu}$, open a window to investigate issues related to integrability of the 2-dimensional sigma model. It is known that kappa symmetry of the two-dimensional sigma model does not imply strictly equations of motion of 10-dimensional supergravity, but leads to a more general setup. This has been elaborated in a series of paper (see e.g., [78–81]) and is usually referred to as the generalized supergravity. In this theory the degree of freedom represented by the dilaton and its derivative is replaced by a set of vectors (X^μ, Z^μ, I^μ), and the equations of motion of generalized supergravity drop to the conventional ones upon $X_\mu = \partial_\mu \varphi$. The most amazing feature of this theory is that (at least some) solutions of its equations of motion can be obtained as integrable deformations of backgrounds of the conventional sugra.

Integrability of the two-dimensional sigma model on the $AdS_5 \times \mathbb{S}^5$ background has been shown in [82] by explicit construction of the corresponding Lax pair. Such sigma model gain the expected interpretation as Type II string theory living on the $AdS_5 \times \mathbb{S}^5$ background supported by non-vanishing flux of 5-form. In [83] a deformed version of the $AdS_5 \times \mathbb{S}^5$ sigma-model has been considered, which appeared to be also integrable for the so-called η-deformations. The η-deformed sigma model is very peculiar when understood as theory of string on a classical background. In this case, the corresponding background, which is commonly referred to as $(AdS_5 \times \mathbb{S}^5)_\eta$, does not satisfy equations of motion of 10-dimensional supergravity [84]. However, it still defines a proper background for string propagation due to Weyl invariance, and moreover the corresponding model is integrable. Such a deformed background was shown to satisfy equations of motion of generalized supergravity [81,85,86].

At the level of field theory, generalized supergravity can be obtained from the conventional Type II supergravity by considering equations of motion for background with one isometry direction x^* and with dilaton, linearly depending on the isometry direction $\varphi = \varphi_0 + ax^*$. T-dualizing solutions along x^* by formally applying Buscher rules one obtains solutions of generalized supergravity. In Double Field Theory such procedure is reflected by just changing the role x^* as a geometric coordinate to become non-geometric, as has been shown explicitly in [86,87]. Alternatively, one is able to reproduce generalized supergravity equations of motion by generalized Scherk-Schwarz reduction of exceptional field theory with twist matrices depending on one dual coordinate. This was explicitly shown for the E_6 ExFT in [88]; however, similar procedure can be performed for ExFTs with any U-duality group.

In addition to the nice incorporation of both generalized and normal supergravities in a single picture, DFT provides a simple and straightforward algorithm for generating deformed backgrounds from a given solution. This procedure has been mainly developed in [89–94] and is based on the so-called β-frame of DFT [95]. The basic idea of B vs β frame is that the generalized metric of DFT as an element of the coset space $O(d,d)/O(d) \times O(d)$ can be parametrized in terms of the fields $G_{\mu\nu}, B_{\mu\nu}$ or alternatively $g_{\mu\nu}, \beta^{\mu\nu}$

$$\mathcal{H}_{MN} = \begin{bmatrix} G_{\mu\nu} - B_{\mu\rho} B^\rho{}_\nu & -B_\mu{}^\nu \\ B^\mu{}_\nu & G^{\mu\nu} \end{bmatrix} = \begin{bmatrix} g_{\mu\nu} & -b_\mu{}^\nu \\ b^\mu{}_\nu & g^{\mu\nu} - \beta^{\mu\rho} \beta_\rho{}^\nu \end{bmatrix}. \tag{93}$$

This is equivalent to choosing the generalized vielbein to be of lower-triangular or upper-triangular form. In principle, one may keep both $B_{\mu\nu}$ and $\beta^{\mu\nu}$ fields in the metric; however this will duplicate gauge degrees of freedom and require either an additional constraint or change in the status of one of the fields. When substituted into the full action of DFT

$$S_{HHZ} = \int dx d\tilde{x} e^{-2d} \left(\frac{1}{8} \mathcal{H}^{MN} \partial_M \mathcal{H}^{KL} \partial_N \mathcal{H}_{KL} - \frac{1}{2} \mathcal{H}^{KL} \partial_L \mathcal{H}^{MN} \partial_N \mathcal{H}_{KM} - 2 \partial_M d \partial_N \mathcal{H}^{MN} + 4 \mathcal{H}^{MN} \partial_M d \partial_N d \right). \tag{94}$$

the generalized metric in β-frame provides the action of the so-called β-supergravity. This theory has been initially developed to address non-geometric fluxes Q and R, which are more naturally written in terms of the bivector field, as in the previous subsection.

The matrix equality (93) can be written in terms of the fundamental fields in the following simple form

$$(G + B)^{-1} = g^{-1} + \beta. \tag{95}$$

In the above one immediately recognizes the open-closed string map [96] that relates backgrounds as seen by the closed string to that seen by the open string. In this interpretation the field $\beta^{\mu\nu}$ on the RHS is understood as the non-commutativity parameter, usually denoted $\Theta^{\mu\nu}$.

The relation (93) can be understood as a deformation of a solution given by the metric background $g_{\mu\nu}$ by a coordinate-dependent parameter $\beta^{\mu\nu}$ (see Figure 5). One starts with a background with vanishing B-field, parametrized only by metric degrees of freedom $g_{\mu\nu}$ and the dilaton Φ. Since $B_{\mu\nu} = 0$, the corresponding generalized metric can be understood in either B or β-frame. Consider now a deformed generalized metric, which contains the metric $g_{\mu\nu}$ that solves conventional equations of motion, and the field β. Then, written in the B-frame this generalized metric encodes a solution of equations of motion of the conventional supergravity. Equivalently, this implies that the generalized metric solves equations of motion of DFT given the section constraint is satisfied. In the β-frame, given $g_{\mu\nu}$ is a solution, these equations of motion impose constraints on the bi-vector field $\beta^{\mu\nu}$ for the deformation to give a solution. In [93] these constraints have been obtained explicitly from equations of motion of β-supergravity of [97]. Given a general solution of these equations one is guaranteed to recover solutions of generalized supergravity after deformation, and in the special case, when $\partial_\mu \beta^{\mu\nu} = 0$, these boil down to solutions of the conventional supergravity.

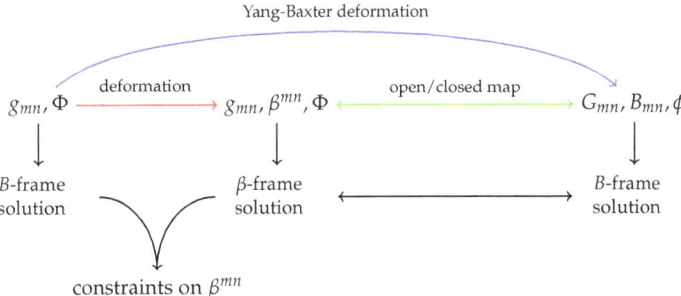

Figure 5. Relationships between the relevant theories and their solutions. b-frame refers to the standard supergravity (possibly generalised), while β-frame is the theory of [98]. Yang-Baxter deformation acts within usual supergravity, but we interpret it as a composition of the open/closed string map with a deformation by β^{mn}. This leads to the constraints for β^{mn} (essentially the CYBE) arising from supergravity field equations.

Of special interest are backgrounds with a set of isometries encoded by Killing vectors $k_a = k_a{}^\mu \partial_\mu$, which form an algebra upon

$$[k_a, k_b] = f_{ab}{}^c k_c. \tag{96}$$

For these the deformation parameter can be chosen in the bi-Killing form

$$\beta^{\mu\nu} = k_a{}^\mu k_b{}^\nu r^{ab}, \tag{97}$$

where r^{ab} is a constant antisymmetric matrix. Such deformed backgrounds satisfy equations of motion of supergavity if the r-matrix satisfied classical Yang-Baxter equation, which is the semi-classical limit of the quantum Yang-Baxter equation

$$r^{[a|b|} r^{c|d|} f_{bd}{}^{e]} = 0. \tag{98}$$

A particular example of such deformation is given by the $(AdS_5 \times \mathbb{S}^5)_\eta$-background discussed above, which corresponds to deformations with r-matrix $r^{ab} f_{ab}{}^c = I^c$. The non-vanishing vector I^c is a sign of the generalized supergravity equations of motion, and indeed $\partial_\mu \beta^{\mu\nu} = k_a{}^\nu I_a$.

The constraint imposed on couplings of the two-dimensional sigma-model by Weyl invariance has been uplifted to the full T-duality covariant sigma-model in [94,99] and proper local counter-terms have been constructed. The covariant approach implies a generalization of the bi-vector deformations described above into a full β-shift. This is a local transformation inside DFT, which can be applied to a solution with non-vanishing H-flux. The general procedure is as follows, one starts with a generalized metric $\mathcal{H}_{MN}$ and a set of generalized Killing vectors $K_a{}^M$, the deformation then is given by

$$\mathcal{H}'_{MN} = h_M{}^K \mathcal{H}_{KL} h_N{}^L, \quad h_M{}^N = \delta_M{}^N - 2\eta r^{ab} K_{aM} K_b{}^N, \tag{99}$$

where η is a constant parameter of deformation. The generalized Killing vector should satisfy the algebraic section constraint (cf. that of [64])

$$\eta_{MN} K_a{}^M K_b{}^N = 0. \tag{100}$$

This ensures the matrix $h_M{}^N$ to be a T-duality transformation, i.e., $h \in O(d,d)$. In the usual duality frame one chooses $K_a{}^M = (k_a{}^\mu, 0)$ and the above transformation reduces to a β-deformation.

4. Conclusions and Discussion

In this short review, we have briefly described the structure of exceptional field theories (ExFT's), which provide a (T)U-duality covariant approach to supergravity. These are based on symmetries of toroidally reduced supergravity; however are defined on a general background. From the point of view of ExFT the toroidal background is a maximally symmetric solution preserving all U-duality symmetries. In this sense the approach is similar to the embedding tensor technique, which is used to define gauge supergravity in a covariant and supersymmetry invariant form. Although any particular choice of gauging breaks certain amount of supersymmetry, the formalism itself is completely invariant. Similarly the U-duality covariant approach is transferred to dynamics of branes in both string and M-theory, whose construction has not been covered here.

In the text, we described construction of the field content of exceptional field theories from fields of dimensionally reduced 11-dimensional supergravity, and local and global symmetries of the theories. Various solutions of the section constraint giving Type IIA/B, 11D and lower-dimensional gauged supergravities have been discussed without going deep into technical details. For readers' convenience references for the original works are present.

As a formalism exceptional field theory has found essential number of application, some of which have been described in this review in more details. In particular, we have covered generalized twist reductions of ExFTs, which reproduce lower-dimensional gauged supergravities, description of non-geometric brane backgrounds and an algorithm for generating deformations of supergravity backgrounds based on frame change inside DFT. However, many fascinating applications of the DFT and ExFT formalisms have been left aside. Among these are non-abelian T-dualities in terms of Poisson-Lie transformations inside DFT [100,101]; generating supersymmetric vacua and consistent truncations of supergravity into lower dimensions [102–104] (for review see [105]); compactifications on non-geometric (Calabi-Yau) backgrounds and construction of cosmological models [54,55,106,107].

Funding: This work was funded by the Russian state grant Goszadanie 3.9904.2017/8.9 and by the Foundation for the Advancement of Theoretical Physics and Mathematics "BASIS".

Conflicts of Interest: The author declares no conflict of interest.

References

1. Maldacena, J.M. The Large N limit of superconformal field theories and supergravity. *Adv. Theor. Math. Phys.* **1998**, *2*, 231–252. [CrossRef]
2. Witten, E. String theory dynamics in various dimensions. *Nucl. Phys.* **1995**, *B443*, 85–126. [CrossRef]
3. Obers, N.; Pioline, B. U duality and M theory. *Phys. Rept.* **1999**, *318*, 113–225. [CrossRef]
4. Hull, C. Global aspects of T-duality, gauged sigma models and T-folds. *JHEP* **2007**, *0710*, 057.
5. Bohm, R.; Gunther, H.; Herrmann, C.; Louis, J. Compactification of type IIB string theory on Calabi-Yau threefolds. *Nucl. Phys.* **2000**, *B569*, 229–246. [CrossRef]
6. Morrison, D.R.; Vafa, C. Compactifications of F theory on Calabi-Yau threefolds. 2. *Nucl. Phys.* **1996**, *B476*, 437–469. [CrossRef]
7. Sen, A. F theory and orientifolds. *Nucl. Phys.* **1996**, *B475*, 562–578. [CrossRef]
8. Musaev, E. U-Dualities in Type II String Theories and M-Theory. Ph.D. Thesis, Queen Mary, University of London, London, UK, 2013.
9. Duff, M. Duality rotations in string theory. *Nucl. Phys.* **1990**, *B335*, 610–620. [CrossRef]
10. Duff, M.; Lu, J. Duality rotations in membrane theory. *Nucl. Phys.* **1990**, *B347*, 394–419. [CrossRef]
11. Cremmer, E.; Julia, B.; Lu, H.; Pope, C. Dualization of dualities. 1. *Nucl. Phys.* **1998**, *B523*, 73–144. [CrossRef]
12. Cremmer, E.; Julia, B.; Lu, H.; Pope, C.N. Dualization of dualities. 2. Twisted self-duality of doubled fields, and superdualities. *Nucl. Phys.* **1998**, *B535*, 242–292. [CrossRef]
13. De Wit, B.; Samtleben, H.; Trigiante, M. The Maximal D = 4 supergravities. *JHEP* **2007**, *0706*, 049. [CrossRef]
14. Berman, D.S.; Cederwall, M.; Kleinschmidt, A.; Thompson, D.C. The gauge structure of generalised diffeomorphisms. *JHEP* **2013**, *1301*, 064. [CrossRef]
15. Hull, C.M. A Geometry for non-geometric string backgrounds. *JHEP* **2005**, *10*, 065.
16. Hull, C.M. Doubled geometry and T-folds. *JHEP* **2007**, *0707*, 080. [CrossRef]
17. De Boer, J.; Shigemori, M. Exotic Branes in String Theory. *Phys. Rep.* **2012**, *532*, 65–118. [CrossRef]
18. Shelton, J.; Taylor, W.; Wecht, B. Nongeometric flux compactifications. *JHEP* **2005**, *0510*, 085. [CrossRef]
19. Wecht, B. Lectures on Nongeometric Flux Compactifications. *Class. Quant. Grav.* **2007**, *24*, S773–S794. [CrossRef]
20. Bergshoeff, E.; Kleinschmidt, A.; Musaev, E.T.; Riccioni, F. The different faces of branes in Double Field Theory. *arXiv* **2019**, arXiv:1903.05601.

21. Hohm, O.; Samtleben, H. Exceptional field theory. III. $E_{8(8)}$. *Phys. Rev.* **2014**, *D90*, 066002. [CrossRef]
22. Kleinschmidt, A.; Nicolai, H. Higher spin representations of K(E10). In Proceedings of the International Workshop on Higher Spin Gauge Theories, Singapore, 4–6 November 2015; World Scientific: Singapore, 2017; pp. 25–38. [CrossRef]
23. Bossard, G.; Cederwall, M.; Kleinschmidt, A.; Palmkvist, J.; Samtleben, H. Generalized diffeomorphisms for E_9. *Phys. Rev.* **2017**, *D96*, 106022. [CrossRef]
24. Bossard, G.; Ciceri, F.; Inverso, G.; Kleinschmidt, A.; Samtleben, H. E_9 exceptional field theory. Part I. The potential. *JHEP* **2019**, *03*, 089.
25. Cederwall, M.; Palmkvist, J. L_∞ algebras for extended geometry. *J. Phys. Conf. Ser.* **2019**, *1194*, 012021. [CrossRef]
26. Samtleben, H. Lectures on Gauged Supergravity and Flux Compactifications. *Class. Quant. Grav.* **2008**, *25*, 214002. [CrossRef]
27. Abzalov, A.; Bakhmatov, I.; Musaev, E.T. Exceptional field theory: $SO(5,5)$. *JHEP* **2015**, *06*, 088. [CrossRef]
28. Musaev, E.T. Exceptional field theory: $SL(5)$. *JHEP* **2016**, *02*, 012. [CrossRef]
29. Samtleben, H.; Weidner, M. The Maximal D = 7 supergravities. *Nucl. Phys.* **2005**, *B725*, 383–419. [CrossRef]
30. Hohm, O.; Samtleben, H. Exceptional Field Theory I: $E_{6(6)}$ covariant Form of M-Theory and Type IIB. *Phys. Rev.* **2014**, *D89*, 066016.
31. Godazgar, H.; Godazgar, M.; Hohm, O.; Nicolai, H.; Samtleben, H. Supersymmetric $E_{7(7)}$ Exceptional Field Theory. *JHEP* **2014**, *1409*, 044. [CrossRef]
32. Musaev, E.; Samtleben, H. Fermions and supersymmetry in $E_{6(6)}$ exceptional field theory. *JHEP* **2015**, *1503*, 027. [CrossRef]
33. Hohm, O.; Samtleben, H. Exceptional field theory. II. $E_{7(7)}$. *Phys. Rev.* **2014**, *D89*, 066017. [CrossRef]
34. Scherk, J.; Schwarz, J.H. How to Get Masses from Extra Dimensions. *Nucl. Phys.* **1979**, *B153*, 61–88. [CrossRef]
35. Graña, M.; Marqués, D. Gauged Double Field Theory. *JHEP* **2012**, *1204*, 020. [CrossRef]
36. Berman, D.S.; Musaev, E.T.; Thompson, D.C. Duality Invariant M-theory: Gauged supergravities and Scherk-Schwarz reductions. *JHEP* **2012**, *1210*, 174. [CrossRef]
37. Musaev, E.T. Gauged supergravities in 5 and 6 dimensions from generalised Scherk-Schwarz reductions. *JHEP* **2013**, *1305*, 161. [CrossRef]
38. Baron, W.H. Gaugings from $E_{7(7)}$ extended geometries. *Phys. Rev.* **2015**, *D91*, 024008. [CrossRef]
39. Hohm, O.; Musaev, E.T.; Samtleben, H. $O(d+1,d+1)$ enhanced double field theory. *JHEP* **2017**, *10*, 086. [CrossRef]
40. Nicolai, H.; Samtleben, H. Maximal gauged supergravity in three-dimensions. *Phys. Rev. Lett.* **2001**, *86*, 1686–1689, [CrossRef]
41. De Wit, B.; Samtleben, H.; Trigiante, M. On Lagrangians and gaugings of maximal supergravities. *Nucl. Phys.* **2003**, *B655*, 93–126. [CrossRef]
42. De Wit, B.; Nicolai, H.; Samtleben, H. Gauged supergravities in three-dimensions: A Panoramic overview. *arXiv* **2004**, arXiv:hep-th/0403014.
43. Trigiante, M. Gauged Supergravities. *Phys. Rept.* **2017**, *680*, 1–175. [CrossRef]
44. Le Diffon, A.; Samtleben, H. Supergravities without an Action: Gauging the Trombone. *Nucl. Phys.* **2009**, *B811*, 1–35. [CrossRef]
45. De Wit, B.; Nicolai, H.; Samtleben, H. Gauged Supergravities, Tensor Hierarchies, and M-Theory. *JHEP* **2008**, *0802*, 044. [CrossRef]
46. De Wit, B.; Samtleben, H.; Trigiante, M. The Maximal D = 5 supergravities. *Nucl. Phys.* **2005**, *B716*, 215–247. [CrossRef]
47. Bergshoeff, E.; Samtleben, H.; Sezgin, E. The Gaugings of Maximal D = 6 Supergravity. *JHEP* **2008**, *0803*, 068. [CrossRef]
48. Cremmer, E.; Julia, B. The SO(8) supergravity. *Nucl. Phys.* **1979**, *B159*, 141–212. [CrossRef]
49. De Wit, B.; Nicolai, H. N = 8 Supergravity with Local SO(8) $\times$ SU(8) Invariance. *Phys. Lett.* **1982**, *B108*, 285–290. [CrossRef]

50. Dibitetto, G.; Fernandez-Melgarejo, J.; Marques, D.; Roest, D. Duality orbits of non-geometric fluxes. *Fortsch. Phys.* **2012**, *60*, 1123–1149. [CrossRef]
51. Danielsson, U.; Dibitetto, G. On the distribution of stable de Sitter vacua. *JHEP* **2013**, *03*, 018. [CrossRef]
52. Damian, C.; Loaiza-Brito, O.; Rey, L.; Sabido, M. Slow-Roll Inflation in Non-geometric Flux Compactification. *J. High Energy Phys.* **2013**, *2013*, 109. [CrossRef]
53. Blabäck, J.; Danielsson, U.; Dibitetto, G. Fully stable dS vacua from generalised fluxes. *JHEP* **2013**, *08*, 054. [CrossRef]
54. Hassler, F.; Lust, D.; Massai, S. On Inflation and de Sitter in Non-Geometric String Backgrounds. *arXiv* **2014**, arXiv:1405.2325.
55. Ma, C.T.; Shen, C.M. Cosmological Implications from O(D,D). *Fortsch. Phys.* **2014**, *62*, 921–941. [CrossRef]
56. Blumenhagen, R.; Kors, B.; Lust, D.; Stieberger, S. Four-dimensional String Compactifications with D-Branes, Orientifolds and Fluxes. *Phys. Rept.* **2007**, *445*, 1–193. [CrossRef]
57. Riccioni, F.; West, P.C. E(11)-extended spacetime and gauged supergravities. *JHEP* **2008**, *0802*, 039. [CrossRef]
58. Kleinschmidt, A. Counting supersymmetric branes. *JHEP* **2011**, *10*, 144. [CrossRef]
59. Lombardo, D.M.; Riccioni, F.; Risoli, S. P fluxes and exotic branes. *JHEP* **2016**, *12*, 114. [CrossRef]
60. Gross, D.J.; Perry, M.J. Magnetic Monopoles in Kaluza-Klein Theories. *Nucl. Phys.* **1983**, *B226*, 29–48. [CrossRef]
61. Hull, C. Generalised geometry for M-theory. *JHEP* **2007**, *0707*, 079. [CrossRef]
62. Andriot, D.; Hohm, O.; Larfors, M.; Lust, D.; Patalong, P. Non-Geometric Fluxes in Supergravity and Double Field Theory. *Fortsch. Phys.* **2012**, *60*, 1150–1186. [CrossRef]
63. Chatzistavrakidis, A.; Gautason, F.F.; Moutsopoulos, G.; Zagermann, M. Effective actions of nongeometric five-branes. *Phys. Rev.* **2014**, *D89*, 066004. [CrossRef]
64. Blair, C.D.A.; Musaev, E.T. Five-brane actions in double field theory. *JHEP* **2018**, *03*, 111. [CrossRef]
65. Andriot, D.; Larfors, M.; Lust, D.; Patalong, P. (Non-)commutative closed string on T-dual toroidal backgrounds. *JHEP* **2013**, *06*, 021. [CrossRef]
66. Blair, C.D.A. Non-commutativity and non-associativity of the doubled string in non-geometric backgrounds. *JHEP* **2015**, *06*, 091. [CrossRef]
67. Berman, D.S.; Rudolph, F.J. Branes are Waves and Monopoles. *JHEP* **2015**, *05*, 015. [CrossRef]
68. Bakhmatov, I.; Kleinschmidt, A.; Musaev, E.T. Non-geometric branes are DFT monopoles. *JHEP* **2016**, *10*, 076. [CrossRef]
69. Bakhmatov, I.; Berman, D.; Kleinschmidt, A.; Musaev, E.; Otsuki, R. Exotic branes in Exceptional Field Theory: The SL(5) duality group. *JHEP* **2018**, *08*, 021. [CrossRef]
70. Fernández-Melgarejo, J.J.; Kimura, T.; Sakatani, Y. Weaving the Exotic Web. *JHEP* **2018**, *09*, 072. [CrossRef]
71. Berman, D.S.; Musaev, E.T.; Otsuki, R. Exotic Branes in Exceptional Field Theory: $E_{7(7)}$ and Beyond. *JHEP* **2018**, *12*, 053. [CrossRef]
72. Berman, D.S.; Musaev, E.T.; Otsuki, R. Exotic Branes in M-Theory. In Proceedings of the Dualities and Generalized Geometries, Corfu, Greece, 9–16 September 2018; *arXiv* **2019**, arXiv:hep-th/1903.10247.
73. Jensen, S. The KK-Monopole/NS5-Brane in Doubled Geometry. *JHEP* **2011**, *1107*, 088. [CrossRef]
74. Gregory, R.; Harvey, J.A.; Moore, G.W. Unwinding strings and t duality of Kaluza-Klein and h monopoles. *Adv. Theor. Math. Phys.* **1997**, *1*, 283–297. [CrossRef]
75. Tong, D. NS5-branes, T duality and world sheet instantons. *JHEP* **2002**, *07*, 013. [CrossRef]
76. Harvey, J.A.; Jensen, S. Worldsheet instanton corrections to the Kaluza-Klein monopole. *JHEP* **2005**, *10*, 028. [CrossRef]
77. Kimura, T.; Sasaki, S. Worldsheet instanton corrections to 5_2^2-brane geometry. *JHEP* **2013**, *08*, 126. [CrossRef]
78. Fradkin, E.S.; Tseytlin, A.A. Effective Field Theory from Quantized Strings. *Phys. Lett.* **1985**, *158B*, 316–322. [CrossRef]
79. Callan, C.G., Jr.; Martinec, E.J.; Perry, M.J.; Friedan, D. Strings in Background Fields. *Nucl. Phys.* **1985**, *B262*, 593–609. [CrossRef]

80. Hull, C.M.; Townsend, P.K. Finiteness and Conformal Invariance in Nonlinear σ Models. *Nucl. Phys.* **1986**, *B274*, 349–362. [CrossRef]
81. Wulff, L.; Tseytlin, A.A. Kappa-symmetry of superstring sigma model and generalized 10d supergravity equations. *JHEP* **2016**, *06*, 174. [CrossRef]
82. Bena, I.; Polchinski, J.; Roiban, R. Hidden symmetries of the AdS(5) x S**5 superstring. *Phys. Rev.* **2004**, *D69*, 046002. [CrossRef]
83. Delduc, F.; Magro, M.; Vicedo, B. An integrable deformation of the $AdS_5 x S^5$ superstring action. *Phys. Rev. Lett.* **2014**, *112*, 051601. [CrossRef]
84. Arutyunov, G.; Borsato, R.; Frolov, S. Puzzles of η-deformed $AdS_5 \times S^5$. *JHEP* **2015**, *12*, 049. [CrossRef]
85. Arutyunov, G.; Frolov, S.; Hoare, B.; Roiban, R.; Tseytlin, A.A. Scale invariance of the η-deformed $AdS_5 \times S^5$ superstring, T-duality and modified type II equations. *Nucl. Phys.* **2016**, *B903*, 262–303. [CrossRef]
86. Sakamoto, J.i.; Sakatani, Y.; Yoshida, K. Weyl invariance for generalized supergravity backgrounds from the doubled formalism. *PTEP* **2017**, *2017*, 053B07.
87. Sakatani, Y.; Uehara, S.; Yoshida, K. Generalized gravity from modified DFT. *JHEP* **2017**, *04*, 123. [CrossRef]
88. Baguet, A.; Magro, M.; Samtleben, H. Generalized IIB supergravity from exceptional field theory. *JHEP* **2017**, *03*, 100. [CrossRef]
89. Araujo, T.; Bakhmatov, I.; Colgáin, E.Ó.; Sakamoto, J.I.; Sheikh-Jabbari, M.M.; Yoshida, K. Conformal twists, Yang–Baxter σ-models & holographic noncommutativity. *J. Phys.* **2018**, *A51*, 235401. [CrossRef]
90. Araujo, T.; Ó Colgáin, E.; Sakamoto, J.; Sheikh-Jabbari, M.M.; Yoshida, K. *I* in generalized supergravity. *Eur. Phys. J.* **2017**, *C77*, 739. [CrossRef]
91. Sakamoto, J.I.; Sakatani, Y.; Yoshida, K. Homogeneous Yang-Baxter deformations as generalized diffeomorphisms. *J. Phys.* **2017**, *A50*, 415401. [CrossRef]
92. Bakhmatov, I.; Ó Colgáin, E.; Sheikh-Jabbari, M.M.; Yavartanoo, H. Yang-Baxter Deformations Beyond Coset Spaces (a slick way to do TsT). *JHEP* **2018**, *06*, 161. [CrossRef]
93. Bakhmatov, I.; Musaev, E.T. Classical Yang-Baxter equation from β-supergravity. *JHEP* **2019**, *01*, 140. [CrossRef]
94. Sakamoto, J.I.; Sakatani, Y. Local β-deformations and Yang-Baxter sigma model. *JHEP* **2018**, *06*, 147. [CrossRef]
95. Andriot, D.; Hohm, O.; Larfors, M.; Lust, D.; Patalong, P. A geometric action for non-geometric fluxes. *Phys. Rev. Lett.* **2012**, *108*, 261602. [CrossRef]
96. Seiberg, N.; Witten, E. String theory and noncommutative geometry. *JHEP* **1999**, *9909*, 032. [CrossRef]
97. Andriot, D.; Betz, A. NS-branes, source corrected Bianchi identities, and more on backgrounds with non-geometric fluxes. *JHEP* **2014**, *07*, 059. [CrossRef]
98. Andriot, D.; Betz, A. β-supergravity: A ten-dimensional theory with non-geometric fluxes, and its geometric framework. *JHEP* **2013**, *12*, 083. [CrossRef]
99. Fernández-Melgarejo, J.J.; Sakamoto, J.I.; Sakatani, Y.; Yoshida, K. Weyl Invariance of String Theories in Generalized Supergravity Backgrounds. *Phys. Rev. Lett.* **2019**, *122*, 111602. [CrossRef]
100. Catal-Ozer, A. Non-Abelian T-duality as a Transformation in Double Field Theory. *arXiv* **2019**, arXiv:1904.00362.
101. Hassler, F. Poisson-Lie T-Duality in Double Field Theory. *arXiv* **2017**, arXiv:707.08624.
102. Baguet, A.; Hohm, O.; Samtleben, H. Consistent Type IIB Reductions to Maximal 5D Supergravity. *Phys. Rev.* **2015**, *D92*, 065004. [CrossRef]
103. Malek, E.; Samtleben, H.; Vall Camell, V. Supersymmetric AdS_7 and AdS_6 vacua and their minimal consistent truncations from exceptional field theory. *Phys. Lett.* **2018**, *B786*, 171–179. [CrossRef]
104. Malek, E.; Samtleben, H.; Vall Camell, V. Supersymmetric AdS_7 and AdS_6 vacua and their consistent truncations with vector multiplets. *JHEP* **2019**, *04*, 088. [CrossRef]
105. Hohm, O.; Samtleben, H. The many facets of exceptional field theory. In Proceedings of the Dualities and Generalized Geometries, Corfu, Greece, 9–16 September 2018; *arXiv* **2019**, arXiv: 1905.08312.

106. Blumenhagen, R.; Font, A.; Plauschinn, E. Relating Double Field Theory to the Scalar Potential of N = 2 Gauged Supergravity. *arXiv* **2015**, arXiv:1507.08059.
107. Bosque, P.d.; Hassler, F.; Lust, D. Flux Formulation of DFT on Group Manifolds and Generalized Scherk-Schwarz Compactifications. *arXiv* **2015**, arXiv:1509.04176.

© 2019 by the author. Licensee MDPI, Basel, Switzerland. This article is an open access article distributed under the terms and conditions of the Creative Commons Attribution (CC BY) license (http://creativecommons.org/licenses/by/4.0/).

Article

Topologically Protected Duality on The Boundary of Maxwell-BF Theory

Alberto Blasi [1] and Nicola Maggiore [1,2,*]

[1] Dipartimento di Fisica, Università di Genova, via Dodecaneso 33, I-16146 Genova, Italy
[2] I.N.F.N.—Sezione di Genova, via Dodecaneso 33, I-16146 Genova, Italy
* Correspondence: maggiore@ge.infn.it

Received: 4 June 2019; Accepted: 10 July 2019; Published: 15 July 2019

Abstract: The Maxwell-BF theory with a single-sided planar boundary is considered in Euclidean four-dimensional spacetime. The presence of a boundary breaks the Ward identities, which describe the gauge symmetries of the theory, and, using standard methods of quantum field theory, the most general boundary conditions and a nontrivial current algebra on the boundary are derived. The electromagnetic structure, which characterizes the boundary, is used to identify the three-dimensional degrees of freedom, which turn out to be formed by a scalar field and a vector field, related by a duality relation. The induced three-dimensional theory shows a strong–weak coupling duality, which separates different regimes described by different covariant actions. The role of the Maxwell term in the bulk action is discussed, together with the relevance of the topological nature of the bulk action for the boundary physics.

Keywords: quantum field theory; topological quantum field theory; duality in gauge field theories; boundary quantum field theory

PACS: 03.70.+k Theory of quantized fields; 11.10.-z Field theory

1. Introduction

Topological field theories have been the subject of a thorough investigation in theoretical physics [1–3]. The initial aim was to unveil to what extent they could give hints for better understanding gravity without matter [4], but it was soon recognized that they also had and still have a different role if one adds a boundary [5,6]. Indeed, it is the boundary which plays a physical role and it is on the boundary that the local observables of a new and different physics live. The introduction of a boundary in a field theory was first proposed by Symanzik who introduced a separability ansatz in order to study the Casimir effect of two parallel plates [7]. The method proposed by Symanzik concerns a space divided into a left and a right hand side, and it has been applied to topological field theories of different types [8], obtaining results particularly relevant for the theory of the fractional quantum Hall effect [9] and of the topological insulators in three and four spacetime dimensions [10]. Later on, field theories with a single-sided boundary have also been considered, which also lead to interesting results. The first and most studied model is the Chern–Simons theory in three dimensions [11,12]. Soon after appeared the BF models [13], the generalizations of the Chern–Simons model which can only live in an odd spacetime [14,15] . The topological nature of the Chern–Simons action is that it does not depend on the metric tensor and hence the energy momentum tensor vanishes. Later on, Witten proposed another kind of topological theory where the action is not in the cohomology of the BRS operator and hence the theory has no physical observables [16]. The common denominator to all these models is that they are gauge field theories with the gauge field A_μ as the main actor and where the Maxwell term does not appear since it breaks the topological nature of the model, which seems to be at the basis of duality relations characterizing the boundary degrees of freedom of these models, whose bulk theories are purely topological [17]. Duality relations of this type are known [18,19] to allow to extract fermionic degrees of freedom out of bosonic ones, in a way compatible with the existence of Hall or quantum spin states on the edge of higher-dimensional bulk theories. It is natural to ask the question to what extent this is true, investigating whether fermionizing duality relations hold also on the boundary of non-topological field theories, which would broaden the possible candidates for the theories of fractional quantum Hall effect and of topological insulators. Moreover, the Maxwell coupling is expected to be quite relevant in whatever physics may arise on the boundary and that is why we are studying models where the Maxwell term is included in the bulk action. This has been done for Chern–Simons theory with both double- [20] and single-sided [21] boundary, with significantly different results. We note also that the introduction of a Maxwell term in the Chern–Simons action gives rise to topologically massive theories [22,23] by means of a mechanism which cannot to be replicated in spacetime dimensions other than three. The question also arises in what sense a topological theory with a Maxwell coupling is still topological. We propose here the Maxwell-BF model as a new kind of topological theory in four spacetime dimensions. The model does not fit into the known "topological" classes of quantum field theories since it does depend explicitly on the metric and the Maxwell term makes it cohomologically nontrivial. Nevertheless, the bulk theory has no local observables due to the equations of motion, which enforce the field strength $F_{\mu\nu}$ to vanish, and hence the gauge field A_μ is pure gauge. This model has never been considered before with a boundary and the question is not irrelevant. The fact that the physics of the model live on the boundary is neither intuitive nor immediately deducible as a simple exercise.

The paper is organized as follows. In Section 2, the model with planar boundary is introduced. The boundary conditions and the Ward identities, broken by the presence of the boundary, are derived. A kind of electromagnetic structure is found on the boundary, with Maxwell equations solved by potentials, which will play the role of degrees of freedom for the 3D theory. The identification of electric and magnetic fields makes possible a physical interpretation of the role of the Maxwell term, as deformation of the magnetic field. In Section 3, following standard methods of quantum field theory, the algebra formed by the conserved electromagnetic currents is found, which heavily depends on the Maxwell term in the bulk action. The algebra, written in terms of the 3D potentials, allows for the construction, in Section 4, of the 3D theory, whose symmetries are identified. The holographic contact is realized by means of the equations of motion of the 3D theory, which are required to be compatible with the boundary conditions of the bulk 4D theory. The resulting equation is recognized to be the duality relation which characterizes the existence of fermionic degrees of freedom on the boundary and therefore turns out not to be peculiar of purely topological bulk field theories only. In other words, the fact that the physical properties are the same in the holographic theory, whether the Maxwell term is present or not, clarifies the meaning of topological quantum field theories when a boundary is introduced.

2. The Model: Action, Boundary Conditions and Ward Identities

The action of the 4D Maxwell-BF theory in Euclidean spacetime is

$$S_{bulk} = \int d^4x\, \theta(x_3) \left(k_1\, \epsilon_{\mu\nu\rho\sigma} F_{\mu\nu} B_{\rho\sigma} + k_2\, F_{\mu\nu} F_{\mu\nu} \right), \tag{1}$$

where the presence of the step-function $\theta(x_3)$ restricts the model on the half-space $x_3 \geq 0$, with planar boundary at $x_3 = 0$, $F_{\mu\nu}(x)$ is the electromagnetic tensor for the gauge field $A_\mu(x)$ and $B_{\mu\nu}(x) = -B_{\nu\mu}(x)$ is the rank-2 antisymmetric tensor of the 4D topological BF theory [24,25]. The canonical mass dimensions of the quantum fields are

$$[A] = 1 \,;\, [B] = 2. \tag{2}$$

Finally, k_1 and k_2 are constants that could be reabsorbed by a redefinition of the fields, but we prefer to keep to be able to trace the contributions of the topological BF and Maxwell F^2 term, respectively.

In absence of boundary, i.e., without the θ-function in Equation (1), the action of the Maxwell-BF theory is invariant under the following two symmetries:

$$\begin{aligned} \delta^{(1)} A_\mu &= \partial_\mu \Lambda \\ \delta^{(1)} B_{\mu\nu} &= 0 \end{aligned} \qquad (3)$$

and

$$\begin{aligned} \delta^{(2)} A_\mu &= 0 \\ \delta^{(2)} B_{\mu\nu} &= \partial_\mu \zeta_\nu - \partial_\nu \zeta_\mu \,, \end{aligned} \qquad (4)$$

where $\Lambda(x)$ and $\zeta_\mu(x)$ are gauge parameters. The presence of the boundary in S_{bulk} breaks the $\delta^{(2)}$-invariance, preserving the usual gauge symmetry $\delta^{(1)}$:

$$\delta^{(1)} S_{bulk} = 0 \;;\; \delta^{(2)} S_{bulk} = -4k_2 \int d^4x \, \delta(x_3) \zeta_i \epsilon_{ijk} \partial_j A_k \,, \qquad (5)$$

where Latin indices run from 0 to 2: $i, j, \ldots = \{0, 1, 2\}$.

The total action S_{tot} is composed by four terms

$$S_{tot} = S_{bulk} + S_{gf} + S_{ext} + S_{bd} \,, \qquad (6)$$

where S_{bulk} is given by Equation (1), S_{gf} is the gauge fixing term

$$S_{gf} = \int d^4x \, \theta(x_3) \, (b A_3 + d_i B_{3i}) \,, \qquad (7)$$

and $b(x)$ and $d_i(x)$ are Lagrange multipliers implementing the gauge conditions

$$A_3 = B_{3i} = 0 \,. \qquad (8)$$

In S_{ext}, external sources $J_i(x)$ and $J_{ij}(x)$ are introduced

$$S_{ext} = \int d^4x \, \theta(x_3) \left(J_i A_i + \frac{1}{2} J_{ij} B_{ij} \right) , \qquad (9)$$

by means of which the quantum fields surviving the gauge conditions in Equation (8) can be defined. Finally, S_{bd} is the most general boundary term defined on $x_3 = 0$ compatible with power counting

$$S_{bd} = \int d^4x \, \delta(x_3) \left(a_1 \epsilon_{ijk} A_i \partial_j A_k + a_2 A_i \partial_3 A_i + a_3 \epsilon_{ijk} A_i B_{jk} + \frac{m}{2} A_i A_i \right) , \qquad (10)$$

where a_i and m are constant parameters and the canonical mass assignments in Equation (2) and $[\delta] = 1$ have been used. Notice that

$$\begin{aligned} \int d^4x \, \theta(x_3) \epsilon_{\mu\nu\rho\sigma} F_{\mu\nu} F_{\rho\sigma} &= 2 \int d^4x \, \theta(x_3) \partial_\mu (\epsilon_{\mu\nu\rho\sigma} A_\nu F_{\rho\sigma}) \\ &= -4 \int d^4x \, \delta(x_3) \epsilon_{ijk} A_i \partial_j A_k \,, \end{aligned} \qquad (11)$$

which justifies the fact that we did not introduce in S_{bulk} (Equation (1)) a term $F\tilde{F}$ in favor of its Chern-Simons-like boundary counterpart, identified by the constant a_1 in Equation (10). In addition, we chose to keep 3D covariance on the boundary. A more general, non-covariant boundary term could have been written [26–28].

The equations of motion of the fields A_i and B_{ij} are

$$\frac{\delta S_{tot}}{\delta A_i} = \theta(x_3)[-2k_1\epsilon_{ijk}\partial_3 B_{jk} - 4k_2\partial_3^2 A_i - 4k_2\partial_j^2 A_i + 4k_2\partial_i\partial_j A_j + J_i] \qquad (12)$$
$$+ \delta(x_3)[(a_3 - 2k_1)\epsilon_{ijk}B_{jk} + (a_2 - 4k_2)\partial_3 A_i + 2a_1\epsilon_{ijk}\partial_j A_k + mA_i] = 0$$

and

$$\frac{\delta S_{tot}}{\delta B_{ij}} = \theta(x_3)[4k_1\epsilon_{ijk}\partial_3 A_k + J_{ij}] + \delta(x_3)[2a_3\epsilon_{ijk}A_k] = 0. \qquad (13)$$

From the equations of motion (Equations (12) and (13)) and performing $\lim_{\epsilon \to 0} \int_{-\epsilon}^{+\epsilon} dx_3$, we get the boundary conditions

$$(a_3 - 2k_1)\epsilon_{ijk}B_{jk} + (a_2 - 4k_2)\partial_3 A_i + 2a_1\epsilon_{ijk}\partial_j A_k + mA_i\Big|_{x_3=0} = 0 \qquad (14)$$

$$a_3 A_i|_{x_3=0} = 0. \qquad (15)$$

The equations of motion lead also to the following Ward identities, broken due to the presence of the boundary

$$\int_0^\infty dx_3\, \partial_i J_i = -\partial_i \left(2k_1\epsilon_{ijk}B_{jk} + 4k_2\partial_3 A_i\right)\Big|_{x_3=0} \qquad (16)$$

$$\int_0^\infty dx_3\, \partial_j J_{ij} = (4k_1 - 2a_3)\, \epsilon_{ijk}\partial_j A_k\Big|_{x_3=0}, \qquad (17)$$

where, to obtain Equation (16), the boundary conditions in Equation (14) have been used. On the boundary $x_3 = 0$ and at vanishing external sources, i.e., on the mass shell, the above Ward identities imply

$$\partial_i \left(2k_1\epsilon_{ijk}B_{jk} + 4k_2\partial_3 A_i\right)\Big|_{x_3=0} = 0 \qquad (18)$$

and

$$\epsilon_{ijk}\partial_j A_k\Big|_{x_3=0} = 0. \qquad (19)$$

Notice that, due to Equation (19), the a_1-term in the boundary action S_{bd} in Equation (10) vanishes. Therefore, without loss of generality, we may rule it out

$$a_1 = 0. \qquad (20)$$

Equations (18) and (19) reveal an electromagnetic structure on the boundary $x_3 = 0$, since they suggest to define an "electric" and a "magnetic" field:

$$\mathcal{E}_i \equiv A_i \qquad (21)$$
$$\mathcal{H}_i \equiv k_1\epsilon_{ijk}B_{jk} + 2k_2\partial_3 A_i, \qquad (22)$$

which allow identifying the degrees of freedom on the boundary $x_3 = 0$ as the corresponding electromagnetic potentials. Indeed, Equations (18) and (19) are solved by introducing a 3D vector field $\tilde{\zeta}_i(X)$ and a 3D scalar field $\Phi(X)$, respectively:

$$\sqrt{M}\epsilon_{ijk}\partial_j\tilde{\zeta}_k(X) \equiv 2k_1\epsilon_{ijk}B_{jk}(x) + 4k_2\partial_3 A_i(x)\Big|_{x_3=0} \qquad (23)$$

$$\frac{1}{\sqrt{M}}\partial_i\Phi(X) \equiv A_i(x)|_{x_3=0}, \qquad (24)$$

where a massive scaling parameter M has been introduced in order to make compatible the mass dimensions in Equation (2) of the 4D fields A_μ and $B_{\mu\nu}$ with those of their 3D boundary counterparts Φ and ξ_i [29]. In 3D spacetime dimensions, in fact, vector fields and scalar fields should have the following mass dimensions:

$$[\xi] = [\Phi] = \frac{1}{2}. \tag{25}$$

A comment is in order: It turns out that a particular solution of the general boundary conditions in Equations (14) and (15) exists, which makes the physics independent of the Maxwell term. In fact, the choice

$$a_2 \neq 4k_2 \,; a_3 = 2k_1 \,; m = \text{any} \tag{26}$$

implies, from Equations (14) and (15), (Dirichlet and) Neumann boundary conditions for the gauge field A_i on $x_3 = 0$

$$\partial_3 A_i = 0|_{x_3=0}. \tag{27}$$

Consequently, the Ward identities do not depend on the coupling k_2 of the Maxwell term in the action in Equation (1), and the theory is indistinguishable, under any respect, from the pure topological BF theory with planar boundary. The fact that the Neumann condition for the gauge field A_i, which is a solution of the general boundary conditions in Equations (14) and (15), makes the Maxwell term transparent and the non-topological theory equivalent to a topological one, is the first nontrivial result of this paper. Since the aim of this paper is to study if and how the non-topological Maxwell term has an impact on the physics on the boundary, we proceed from now disregarding the solution in Equation (26). In particular, the boundary condition in Equation (15) is solved by

$$a_3 = 0. \tag{28}$$

The k_2-Maxwell term manifests itself on the r.h.s. of the Ward identity in Equation (16) by means of $\partial_3 A_i|_{x_3=0}$. We observe that, on the boundary $x_3 = 0$, the fields $A_i|_{x_3=0}$ and $\partial_3 A_i|_{x_3=0}$ must be treated as independent dynamical fields [30]. Consequently, we need to couple, on the boundary $x_3 = 0$, an external source $\hat{J}_i$ to $\partial_3 A_i|_{x_3=0}$, as done in [31], where the 3D Maxwell theory with boundary has been studied:

$$S_{ext} \to \hat{S}_{ext} = \int d^4x \left[\theta(x_3) \left(J_i A_i + \frac{1}{2} J_{ij} B_{ij} \right) + \delta(x_3) \hat{J}_i \partial_3 A_i \right]. \tag{29}$$

3. The Boundary Algebra

Differentiating the two Ward identities in Equations (16) and (17) with respect to the external sources $J_i(x)$, $J_{ij}(x)$ and $\hat{J}_i(x)$ and then going at $J = 0$ lead to six algebraic relations. We consider the subalgebra obtained as follows.

Differentiating the Ward identity in Equation (16) with respect to $J_m(x')$, and then going at vanishing external source, we get

$$\partial_m^X \delta^{(3)}(X - X') = \partial_i^X \left(-2k_1 \epsilon_{ijk} \Delta_{A_m B_{jk}}(X', X) - 4k_2 \Delta_{A_m \partial_3 A_i}(X', X) \right). \tag{30}$$

In Equation (30), $\partial_i^X \equiv \frac{\partial}{\partial X_i}$, and the time-ordered propagator between two generic fields $\Phi(X)$ and $\Psi(X)$ is defined on the generating functional of the connected Green functions $Z_c[J]$, as usual, as

$$\Delta_{\Phi\Psi}(X, X') \equiv \left. \frac{\delta^{(2)} Z_c}{\delta J_\Phi(X) \delta J_\Psi(X')} \right|_{J_\Phi = J_\Psi = 0}$$
$$= \theta(x_0 - x_0') \langle \Phi(X) \Psi(X') \rangle + \theta(x_0' - x_0) \langle \Psi(X') \Phi(X) \rangle. \tag{31}$$

Hence, from Equation (30), we have

$$\partial_m^X \delta^{(3)}(X - X') = \delta(x_0 - x'_0)\langle[A_m(X'), 2k_1\widetilde{B}(X) + 4k_2\partial_3 A_0(X)]\rangle$$
$$- 2\theta(x_0 - x'_0)\langle\left[\partial_i^X\left(k_1\epsilon_{ijk}B_{jk} + 2k_2\partial_3 A_i\right)(X)A_m(X')\right]\rangle \quad (32)$$
$$- 2\theta(x'_0 - x_0)\langle\left[A_m(X')\partial_i^X\left(k_1\epsilon_{ijk}B_{jk} + 2k_2\partial_3 A_i\right)(X)\right]\rangle,$$

where we define

$$\widetilde{B}(X) \equiv \epsilon_{\alpha\beta}B_{\alpha\beta}(X). \quad (33)$$

The last two terms on the r.h.s. of Equation (32) vanish on-shell due to Equation (18), so that, at vanishing external sources, we get

$$\partial_m^X \delta^{(3)}(X - X') = \delta(x_0 - x'_0)\langle[A_m(X'), 2k_1\widetilde{B}(X) + 4k_2\partial_3 A_0(X)]\rangle. \quad (34)$$

Following the same steps, differentiating the Ward identity in Equation (16) with respect to the external sources $J_{mn}(x')$ and $\hat{J}_m(x')$, we get, respectively,

$$\delta(x_0 - x'_0)\langle[B_{mn}(X'), k_1\widetilde{B}(X) + 2k_2\partial_3 A_0(X)]\rangle = 0 \quad (35)$$

and

$$\delta(x_0 - x'_0)\langle[k_1\widetilde{B}(X) + 2k_2\partial_3 A_0(X), \partial_3 A_m(X')]\rangle = 0. \quad (36)$$

On the other hand, deriving the Ward identity in Equation (17) with respect to $J_{mn}(x')$, we obtain

$$\delta(x_0 - x'_0)\langle[B_{mn}(X'), k_1\widetilde{B}(X) + 2k_2\partial_3 A_0(X)]\rangle = 0. \quad (37)$$

From the above algebraic relations, we get the following subalgebra formed by equal-time commutators:

$$\langle[A_\alpha(X), 2k_1\widetilde{B}(X') + 4k_2\partial_3 A_0(X')]\rangle_{x_0=x'_0} = \partial_\alpha^X \delta^{(2)}(X - X') \quad (38)$$

$$\langle[A_\alpha(X), A_\beta(X')]\rangle_{x_0=x'_0} = 0 \quad (39)$$

$$\langle\left[2k_1\widetilde{B}(X) + 4k_2\partial_3 A_0(X), 2k_1\widetilde{B}(X') + 4k_2\partial_3 A_0(X')\right]\rangle_{x_0=x'_0} = 0, \quad (40)$$

which, written in terms of the 3D boundary fields $\xi_i(X)$ (Equation (23)) and $\Phi(X)$ (Equation (24)), implies

$$\langle[\Phi(X), \epsilon_{\alpha\beta}\partial_\alpha\xi_\beta(X')]\rangle_{x_0=x'_0} = \delta^{(2)}(X - X') \quad (41)$$
$$\langle[\Phi(X), \Phi(X')]\rangle_{x_0=x'_0} = 0 \quad (42)$$
$$\langle[\epsilon_{\alpha\beta}\partial_\alpha\xi_\beta(X), \epsilon_{\alpha\beta}\partial_\alpha\xi_\beta(X')]\rangle_{x_0=x'_0} = 0. \quad (43)$$

4. The Action Induced on the 3D Boundary

The commutators in Equations (41)–(43) can be interpreted as equal-time canonical commutation relations for the 3D canonically conjugate variables

$$q(X) \equiv \frac{1}{\sqrt{M}}\Phi(X)$$
$$p(X) \equiv \sqrt{M}\epsilon_{\alpha\beta}\partial_\alpha\xi_\beta(X), \quad (44)$$

and this allows us to identify the most general action $S_{3D}[\Phi, \xi]$ induced on the planar boundary $x_3 = 0$ of the 4D Maxwell-BF theory, which must display the following features:

1. The 3D action $S_{3D}[\Phi, \xi]$ must be a local integrated functional of the 3D fields in Equations (23) and (24), with canonical mass dimension equal to three.
2. The 3D Lagrangian $\mathcal{L}_{3D}[\Phi, \xi]$ must be such that the relation between the canonically conjugate variables in Equation (44) holds true:

$$p(X) = \frac{\partial \mathcal{L}_{3D}}{\partial \dot{q}(X)}, \tag{45}$$

   which implies that the Lagrangian $\mathcal{L}_{3D}[\Phi, \xi]$ must contain time derivatives only in the term $p\dot{q}$.
3. The action $S_{3D}[\Phi, \xi]$ must display the two symmetries, which leave invariant the definitions in Equation (23) and (24):

   (a) gauge symmetry
   $$\delta_{gauge} \xi_i = \partial_i \Lambda \tag{46}$$

   (b) shift symmetry
   $$\delta_{shift} \Phi = \text{constant}. \tag{47}$$

The most general action satisfying the above requests is

$$S_{3D}[\Phi, \xi] = \int d^3 X \left[c_1 (\epsilon_{\alpha\beta} F_{\alpha\beta})(\partial_0 \Phi) + c_2 F_{\alpha\beta} F_{\alpha\beta} + c_3 \partial_\alpha \Phi \partial_\alpha \Phi \right], \tag{48}$$

where $F_{\alpha\beta} = \partial_\alpha \xi_\beta - \partial_\beta \xi_\alpha$ and c_i, $i = 1, 2, 3$ are constants. From the action $S_{3D}[\Phi, \xi]$, we get the equations of motion

$$\frac{\delta S_{3D}}{\delta \Phi} = -2\partial_\alpha (c_1 \epsilon_{\alpha\beta} \partial_0 \xi_\beta + c_3 \partial_\alpha \Phi) = 0 \tag{49}$$

$$\frac{\delta S_{3D}}{\delta \xi_\alpha} = 2\partial_\beta (c_1 \epsilon_{\alpha\beta} \partial_0 \Phi + 2c_2 F_{\alpha\beta}) = 0. \tag{50}$$

5. Holographic Constraint and Duality

The equations of motion (Equations (49) and (50)) of the scalar-vector 3D action $S_{3D}[\Phi, \xi]$ (Equation (48)) must be compatible with the boundary conditions in Equations (14) and (15) of the 4D Maxwell-BF theory on the planar boundary $x_3 = 0$. To make this holographic contact [32], the boundary condition in Equation (14) written in terms of the boundary degrees of freedom ξ_i (Equation (23)) and Φ (Equation (24)) is

$$\epsilon_{ijk} \partial_j \xi_k - \kappa \partial_i \Phi = 0, \tag{51}$$

where, besides Equations (20) and (28), we ch00se

$$a_2 = 0, \tag{52}$$

and we define the dimensionless normalized mass parameter

$$\kappa \equiv \frac{m}{M}. \tag{53}$$

We recognize in Equation (51) the duality relation found in [33], which extracts fermionic degrees of freedom from bosonic ones [18,19,34]. We come back to this point below. Here, it appears as the unique boundary condition that relates the 4D Maxwell-BF theory with boundary and its holographic 3D counterpart(s), as explicitly shown below.

Notice that the three components $i = \{0, \alpha\}$ of Equation (51) are

$$\epsilon_{\alpha\beta}F_{\alpha\beta} - 2\kappa\partial_0\Phi = 0 \quad (54)$$
$$\epsilon_{\alpha\beta}\partial_0\xi_\beta - \epsilon_{\alpha\beta}\partial_\beta\xi_0 + \kappa\partial_\alpha\Phi = 0, \quad (55)$$

which are compatible with Equations (49) and (50) if

$$c_2 = -\frac{1}{2\kappa}c_1 \; ; c_3 = \kappa c_1 \quad (56)$$

and if the temporal gauge choice for the 3D gauge field ξ_i is imposed

$$\xi_0 = 0. \quad (57)$$

6. Summary of Results and Discussion

When a boundary is introduced in a quantum field theory, a crucial role is played by the boundary term, which in the case studied in this paper is represented by S_{bd} (Equation (10)), which depends by a number of constant parameters which need to be fine tuned in order to determine the holographic theory induced on the boundary. For the 4D Maxwell-BF theory, the boundary term finally reduces to

$$S_{bd} = \frac{m}{2} \int d^4x\, \delta(x_3) A_i A_i, \quad (58)$$

which depends on one massive parameter m only.

In presence of a boundary, the question naturally arises of which boundary conditions should be imposed. The procedure described in this paper leads to the following boundary condition compatible with the holographic construction:

$$2k_1 \epsilon_{ijk} B_{jk} + 4k_2 \partial_3 A_i - m A_i \Big|_{x_3=0} = 0, \quad (59)$$

which involves both the k_1-BF and the k_2-Maxwell terms. Notice that it depends on one parameter only (m). It is of a nonstandard type, since it does not fall into the usual Dirichlet, Neumann or Robin boundary conditions on each field, separately. On the contrary, it mixes both the fields, and the effect of the Maxwell term is to introduce a dependence on the transverse component of the gauge field with respect to the planar boundary, which is independent of the longitudinal components, and consequently must be treated as an independent dynamical variable on the boundary. As we remarked, an unexpected consequence of the boundary conditions in Equations (14) and (15) is that they can be solved by the set of parameters in Equation (26), which implies, in particular, Neumann boundary condition for the gauge field A_μ. This corresponds to eliminating any dependence from k_2, i.e., from the Maxwell term, in the physics on the boundary, represented by the current algebra as well as the boundary conditions themselves.

The boundary breaks all the invariances of the unbounded theory: translations, parity and gauge symmetries. The first consequence concerns the choice of the gauge conditions, implemented by the gauge fixing term S_{gf} in Equation (7), which does not need to be covariant. With deeper consequences, the Ward identities describing the Ward identities are broken by boundary terms:

$$\int_0^\infty dx_3\, \partial_i J_i = -\partial_i \left(2k_1 \epsilon_{ijk} B_{jk} + 4k_2 \partial_3 A_i \right) \Big|_{x_3=0} \quad (60)$$

$$\int_0^\infty dx_3\, \partial_j J_{ij} = 4k_1\, \epsilon_{ijk} \partial_j A_k \Big|_{x_3=0}. \quad (61)$$

At $J = 0$, i.e., going on-shell, the vanishing of the boundary breakings lead to define on $x_3 = 0$ an "electric" field and a "magnetic" field, and the corresponding potentials:

$$\mathcal{E}_i \equiv A_i \propto \partial_i \Phi \qquad (62)$$

$$\mathcal{H}_i \equiv k_1 \epsilon_{ijk} B_{jk} + 2k_2 \partial_3 A_i \propto \epsilon_{ijk} \partial_j \xi_k , \qquad (63)$$

which allow physically interpreting the contribution of the Maxwell term as a deformation of the magnetic field on the boundary. Without Maxwell term, i.e., at $k_2 = 0$, the fields A_i and B_{ij} are interpreted as electric and magnetic fields on the boundary, respectively. Their transverse component do not enter in the game and could safely be eliminated by Neumann boundary conditions. In the presence of the Maxwell term, this is no longer true for the gauge field A, for which both the longitudinal and transverse components are physically important: the former as electric field, and the latter as deformation of the main magnetic field represented by the dual of the B-field. The potentials corresponding to the boundary electric and magnetic fields (Equations (62) and (63)) are the degrees of freedom by means of which the holographic 3D theory is constructed: a vector field ξ_i and a scalar field Φ.

The algebra obtained from the broken Ward identities in Equations (60) and (61) by differentiating them with respect to the external sources J, written in terms of the boundary degrees of freedom is

$$\langle [\Phi(X), \epsilon_{\alpha\beta}\partial_\alpha \xi_\beta(X')] \rangle_{x_0 = x'_0} = \delta^{(2)}(X - X') \qquad (64)$$

$$\langle [\Phi(X), \Phi(X')] \rangle_{x_0 = x'_0} = 0 \qquad (65)$$

$$\langle [\epsilon_{\alpha\beta}\partial_\alpha \xi_\beta(X), \epsilon_{\alpha\beta}\partial_\alpha \xi_\beta(X')] \rangle_{x_0 = x'_0} = 0 , \qquad (66)$$

which can be seen as equal-time canonical commutation relations between canonical variables $q(X)$ and $p(X)$.

Once the canonical variables are identified, the corresponding action is found to be

$$S_{3D}[\Phi, \xi] = c_1 \int d^3 X \left[(\epsilon_{\alpha\beta} F_{\alpha\beta})(\partial_0 \Phi) - \frac{1}{2\kappa} F_{\alpha\beta} F_{\alpha\beta} + \kappa \partial_\alpha \Phi \partial_\alpha \Phi \right], \qquad (67)$$

which is the most general 3D local integrated functional built with a scalar and a vector field, respecting power counting and invariant under the gauge (Equation (46)) and shift (Equation (47)) symmetries, and, most importantly, whose equations of motion are compatible with the boundary conditions in Equation (59):

$$\epsilon_{ijk} F_{jk} - 2\kappa \partial_i \Phi = 0 . \qquad (68)$$

Equation (68) coincides with the duality relation between a scalar and a vector field which has been invoked in [18,19,33] as the main tool for the mechanism of fermionization of bosonic degrees of freedom. In other words, the effective dynamical variables living on the boundary of the 4D Maxwell-BF are fermionic. This feature is crucial for the interpretation of the boundary degrees of freedom as the edge states of the 3D topological insulators [35,36]. The new fact that we are recovering here is that this property, which has been related to the topological character of the bulk theory [17], indeed also holds for a non-topological theory such as Maxwell-BF theory. We comment in more detail on this point below. The duality relation in Equation (68) has also more field theoretical consequences. Thanks to Equation (68), indeed, the action in Equation (67) is covariant, despite the appearances. In fact, the scalar field Φ can be eliminated from the action through Equation (68), and we find the 3D Maxwell theory

$$S_{Max}[\xi] = \frac{c_1}{2\kappa} \int d^3 X \, F_{ij} F_{ij} . \qquad (69)$$

Alternatively, the duality relation in Equation (68) allows us to trade the vector for the scalar field

$$S_{KK}[\Phi] = \kappa c_1 \int d^3X \, \partial_i \Phi \partial_i \Phi \,. \tag{70}$$

The scalar–vector 3D theory described by Equation (67), the Maxwell action in Equation (69) and the massless Klein–Gordon action in Equation (70) are all holographic counterparts of the 4D Maxwell-BF theory with planar boundary. The two actions in Equations (69) and (70) are related by the duality relation in Equation (51), which may be written in a way to emphasize its strong–weak coupling aspect:

$$\partial_i \Phi \leftrightarrow \epsilon_{ijk} F_{jk} \,\cup\, \kappa \leftrightarrow \frac{1}{\kappa} \,. \tag{71}$$

In this form, it is apparent that the coupling κ in Equation (53) governs the regimes where one type of action dominates with respect to the other: at strong coupling (very large κ), the dominating term is the Maxwell one, while, at weak coupling (very small κ), it is the massless scalar action which dominates. At intermediate regimes, both the degrees of freedom are present, and the relevant action is Equation (67), whose degrees of freedom are fermionic, due to Equation (68), as remarked above. It is this intermediate regime which is relevant for the topological insulators, for which the same action in Equation (67) has been proposed in [35]. Surprisingly, the order parameter κ, which distinguishes the various regimes, is directly related through Equation (53) to the only effective parameter m, which survives in the boundary term in Equation (58), and hence plays a much more crucial role than one might expect at first sight.

We conclude this paper with a remark concerning the effect of the presence of the Maxwell term in the bulk action in Equation (1) together with a general consideration on the topological nature of quantum field theories. The main feature of topological quantum field theories is the lack of local observables, the only observables being global geometrical properties of the manifolds on which they are built. Technically, this means that the local cohomology of the BRS operator is empty. Examples of topological quantum field theories are the 3D Chern–Simons theory and BF theories, which may be defined in any spacetime dimensions. The 4D Maxwell-BF theory studied in this paper is not topological, since its action depends on the metric through the Maxwell term, and its cohomological structure is nontrivial. Nevertheless, the bulk action has no local observables since the equations of motion enforce the vanishing of the $F_{\mu\nu}$ tensor; thus, we are looking at a different class of "topological" theories where the boundary is expected to carry all the physical information. If a boundary is introduced, as above, this nontriviality reflects in the algebra derived from the Ward identities in Equations (60) and (61), which is rather complicated. We write down the four relations (out of six) as Equations (34)–(37), where the dependence on the Maxwell term is highlighted by the coefficient k_2. In addition, the boundary condition in Equation (59) explicitly depends on the Maxwell term, which physically results in a perturbation of the magnetic field $\mathcal{H}$ (Equation (63)). On the other hand, when constructing the holographic scalar-vector 3D theory, the presence of the Maxwell term is buried in the definition of the vector potential $\tilde{\xi}_i$ in Equation (63), and both 3D actions (in their duality-related representations, Equations (67), (69) and (70), and the duality relation, Equation (68)) are the same as in the case of pure BF with boundary. In other words, from the holographic point of view, Maxwell-BF theory with boundary is indistinguishable from the pure topological BF theory, as if holography would protect the topological character of bulk theories.

Funding: This research received no external funding

Author Contributions: A.B. and N.M. equally contributed to all phases of preparation of this article.

Conflicts of Interest: The authors declare no conflict of interest.

References

1. Witten, E. Supersymmetry and Morse theory. *J. Diff. Geom.* **1982**, *17*, 661. [CrossRef]
2. Witten, E. Topological Quantum Field Theory. *Commun. Math. Phys.* **1988**, *117*, 353. [CrossRef]
3. Atiyah, M. Topological quantum field theories. *Inst. Hautes Etudes Sci. Publ. Math.* **1989**, *68*, 175. [CrossRef]
4. Witten, E. (2+1)-Dimensional Gravity as an Exactly Soluble System. *Nuclear Phys. B* **1988**, *311*, 46. [CrossRef]
5. Moore, G.W.; Seiberg, N. Taming the Conformal Zoo. *Phys. Lett. B* **1989**, *220*, 422. [CrossRef]
6. Cattaneo, A.S.; Mnev, P.; Reshetikhin, N. Classical and quantum Lagrangian field theories with boundary. *PoS CORFU* **2011**, *2011*, 044. [CrossRef]
7. Symanzik, K. Schrodinger Representation and Casimir Effect in Renormalizable Quantum Field Theory. *Nuclear Phys. B* **1981**, *190*, 1. [CrossRef]
8. Amoretti, A.; Blasi, A.; Maggiore, N.; Magnoli, N. Three-dimensional dynamics of four-dimensional topological BF theory with boundary. *New J. Phys.* **2012**, *14*, 113014. [CrossRef]
9. Blasi, A.; Ferraro, D.; Maggiore, N.; Magnoli, N.; Sassetti, M.; Symanzik's Method Applied To The Fractional Quantum Hall Edge States. *Ann. Phys.* **2008**, *17*, 885. [CrossRef]
10. Amoretti, A.; Braggio, A.; Caruso, G.; Maggiore, N.; Magnoli, N. Introduction of a boundary in topological field theories. *Phys. Rev. D* **2014**, *90*, 125006. [CrossRef]
11. Chern, S.S.; Simons, J. Characteristic forms and geometric invariants. *Ann. Math.* **1974**, *99*, 48. [CrossRef]
12. Blasi, A.; Collina, R. The Chern-Simons model with boundary: A Cohomological approach. *Int. J. Mod. Phys. A* **1992**, *7*, 3083. [CrossRef]
13. Horowitz, G.T. Exactly Soluble Diffeomorphism Invariant Theories. *Commun. Math. Phys.* **1989**, *125*, 417. [CrossRef]
14. Karlhede, A.; Rocek, M. Topological Quantum Field Theories in Arbitrary Dimensions. *Phys. Lett. B* **1989**, *224*, 58. [CrossRef]
15. Blasi, A.; Maggiore, N.; Montobbio, M. Noncommutative two dimensional BF model. *Nuclear Phys. B* **2006**, *740*, 281. [CrossRef]
16. Birmingham, D.; Blau, M.; Rakowski, M.; Thompson, G. Topological field theory. *Phys. Rept.* **1991**, *209*, 129. [CrossRef]
17. Amoretti, A.; Braggio, A.; Caruso, G.; Maggiore, N.; Magnoli, N. Holography in flat spacetime: 4D theories and electromagnetic duality on the border. *J. High Energy Phys.* **2014**, *1404*, 142. [CrossRef]
18. Aratyn, H. A Bose Representation For The Massless Dirac Field In Four-dimensions. *Nuclear Phys. B* **1983**, *227*, 172. [CrossRef]
19. Aratyn, H. Fermions From Bosons In (2+1)-dimensions. *Phys. Rev. D* **1983**, *28*, 2016. [CrossRef]
20. Blasi, A.; Maggiore, N.; Magnoli, N.; Storace, S. Maxwell-Chern-Simons Theory with Boundary. *Class. Quant. Grav.* **2010**, *27*, 165018. [CrossRef]
21. Maggiore, N. Holographic reduction of Maxwell-Chern-Simons theory. *Eur. Phys. J. Plus* **2018**, *133*, 281. [CrossRef]
22. Deser, S.; Jackiw, R.; Templeton, S. Topologically Massive Gauge Theories. *Ann. Phys.* **1982**, *140*, 372–411. Erratum in: **2000**, *281*, 409. Erratum in: **1988**, *185*, 406. [CrossRef]
23. Deser, S.; Jackiw, R.; Templeton, S. Three-Dimensional Massive Gauge Theories. *Phys. Rev. Lett.* **1982**, *48*, 975. [CrossRef]
24. Blau, M.; Thompson, G. A New Class of Topological Field Theories and the Ray-singer Torsion. *Phys. Lett. B* **1989**, *228*, 64. [CrossRef]
25. Blau, M.; Thompson, G. Topological Gauge Theories of Antisymmetric Tensor Fields. *Ann. Phys.* **1991**, *205*, 130 [CrossRef]
26. Maggiore, N. From Chern–Simons to Tomonaga–Luttinger. *Int. J. Mod. Phys. A* **2018**, *33*, 1850013. [CrossRef]
27. Blasi, A.; Maggiore, N. Massive deformations of rank-2 symmetric tensor theory (a.k.a. BRS characterization of Fierz–Pauli massive gravity). *Class. Quant. Grav.* **2017**, *34*, 015005. [CrossRef]
28. Blasi, A.; Maggiore, N. Massive gravity and Fierz-Pauli theory. *Eur. Phys. J. C* **2017**, *77*, 614. [CrossRef]
29. Blasi, A.; Maggiore, N.; Montobbio, M. Instabilities of noncommutative two dimensional bf model. *Mod. Phys. Lett. A* **2005**, *20*, 2119. [CrossRef]
30. Karabali, D.; Nair, V.P. Boundary Conditions as Dynamical Fields. *Phys. Rev. D* **2015**, *92*, 125003. [CrossRef]

31. Maggiore, N. Conserved chiral currents on the boundary of 3D Maxwell theory. *J. Phys. A* **2019**, *52*, 115401. [CrossRef]
32. Amoretti, A.; Braggio, A.; Maggiore, N.; Magnoli, N. Thermo-electric transport in gauge/gravity models. *Adv. Phys. X* **2017**, *2*, 409. [CrossRef]
33. Amoretti, A.; Blasi, A.; Caruso, G.; Maggiore, N.; Magnoli, N. Duality and Dimensional Reduction of 5D BF Theory. *Eur. Phys. J. C* **2013**, *73*, 2461. [CrossRef]
34. Amoretti, A.; Braggio, A.; Caruso, G.; Maggiore, N.; Magnoli, N. 3+1D Massless Weyl spinors from bosonic scalar-tensor duality. *Adv. High Energy Phys.* **2014**, *2014*, 635286. [CrossRef]
35. Cho, G.Y.; Moore, J.E. Topological BF field theory description of topological insulators. *Ann. Phys.* **2011**, *326*, 1515. [CrossRef]
36. Blasi, A.; Braggio, A.; Carrega, M.; Ferraro, D.; Maggiore, N.; Magnoli, N. Non-Abelian BF theory for 2+1 dimensional topological states of matter. *New J. Phys.* **2012**, *14*, 013060. [CrossRef]

© 2019 by the authors. Licensee MDPI, Basel, Switzerland. This article is an open access article distributed under the terms and conditions of the Creative Commons Attribution (CC BY) license (http://creativecommons.org/licenses/by/4.0/).

Article

A Conformally Invariant Derivation of Average Electromagnetic Helicity

Ivan Fernandez-Corbaton

Institute of Nanotechnology, Karlsruhe Institute of Technology, 76021 Karlsruhe, Germany;
ivan.fernandez-corbaton@kit.edu

Received: 14 October 2019; Accepted: 8 November 2019; Published: 19 November 2019

Abstract: The average helicity of a given electromagnetic field measures the difference between the number of left- and right-handed photons contained in the field. Here, the average helicity is derived using the conformally invariant inner product for Maxwell fields. Several equivalent integral expressions in momentum space, in $(\mathbf{r}, t)$ space, and in the time-harmonic $(\mathbf{r}, \omega)$ space are obtained, featuring Riemann–Silberstein-like fields and potentials. The time-harmonic expressions can be directly evaluated using the outputs of common numerical solvers of Maxwell equations. The results are shown to be equivalent to the well-known volume integral for the average helicity, featuring the electric and magnetic fields and potentials.

Keywords: electromagnetic helicity; conformal symmetry

The helicity of electromagnetic fields has received research attention since at least the 1960s [1–8]. Recently, the topic has picked up a considerable pace, partly because of the relevance of helicity in chiral light–matter interactions [9–24]. In this context, one of the most basic quantities is the average value of electromagnetic helicity for a given free electromagnetic field, which can be interpreted as the pseudo-scalar that measures the difference between the number of left- and right-handed polarized photons contained in the field [1,2]. The most common expression for average helicity is [1–3,6–8,11–15,17,21–24]:

$$\frac{1}{2} \int_{\mathbb{R}^3} d\mathbf{r}\, \mathcal{B}(\mathbf{r}, t) \cdot \mathcal{A}(\mathbf{r}, t) - \mathcal{E}(\mathbf{r}, t) \cdot \mathcal{C}(\mathbf{r}, t),$$

where $\mathcal{E}(\mathbf{r}, t)$ [$\mathcal{C}(\mathbf{r}, t)$] and $\mathcal{B}(\mathbf{r}, t)$ [$\mathcal{A}(\mathbf{r}, t)$] are the real-valued electric and magnetic fields[potentials], respectively. The use of two potentials is a common strategy in this context [2,8,11–13,17,25]. In particular, it allows the obtaining of an integrand which is local in $\mathbf{r}$. The above equation has been arrived at in several ways. This article contains a different one.

In this article, several different integral expressions for the average helicity of a given electromagnetic field are obtained starting from the inner product that ensures invariance of the result under the largest symmetry group of Maxwell equations: The conformal group. We show that the result from the derived expressions coincides with that of the above integral. In our approach, all the fields are complex because only positive frequencies are included. The advantages of this choice regarding the treatment of helicity with Riemann-Silberstein-like fields and their corresponding potentials are discussed. Integral expressions in momentum space, $(\mathbf{r}, t)$ space, and $(\mathbf{r}, \omega)$ space are obtained. The numerical evaluation of the time-harmonic $(\mathbf{r}, \omega)$ expressions, whose integrands are local in $\mathbf{r}$, can be conveniently performed using common numerical Maxwell solvers. The formalism is used to obtain expressions for the computation of molecular circular dichroism in Appendix C.

We start by writing down the conformally invariant form of the average helicity and showing that the result coincides with the common definition. To such end, we consider $\mathbb{M}$, the vector space of finite-energy solutions to Maxwell's equations in free space. We denote vectors in such space by kets such as $|\mathcal{F}\rangle$, which represent particular electromagnetic field solutions. We are interested in the average helicity of the field. The key ingredient for the definition of average properties is an inner product between two vectors $|\mathcal{F}\rangle$ and $|\mathcal{G}\rangle$, denoted $\langle\mathcal{G}|\mathcal{F}\rangle$, which endows $\mathbb{M}$ with the structure of a Hilbert space. Properties such as energy, linear momentum, helicity, etc...., are represented by Hermitian operators which map elements of $\mathbb{M}$ back onto itself. Then, for a given field $|\mathcal{F}\rangle$, the average value of a given property represented by a Hermitian operator Γ is the quadratic form

$$\langle\mathcal{F}|\Gamma|\mathcal{F}\rangle, \tag{1}$$

that is, the projection of the vector $|\mathcal{F}\rangle$ onto the vector $\Gamma|\mathcal{F}\rangle$. Since Γ is Hermitian, $\langle\mathcal{F}|\Gamma|\mathcal{F}\rangle$ must be a real number.

The crucial question of *which inner product* to choose was settled by Gross by requiring the inner product to be invariant under the conformal group [26]. That is: Given any two solutions $|\mathcal{F}\rangle$ and $|\mathcal{G}\rangle$, and their corresponding transformed versions under any transformation in the conformal group, $|\tilde{\mathcal{F}}\rangle$ and $|\tilde{\mathcal{G}}\rangle$, the inner product must be so that $\langle\mathcal{F}|\mathcal{G}\rangle = \langle\tilde{\mathcal{F}}|\tilde{\mathcal{G}}\rangle$. Gross showed in Ref. [26] that this requirement essentially determines the exact expression of the inner product, which we will use later. The conformal group includes space–time translations, spatial rotations, and Lorentz boosts, which together form the Poincaré group, plus space–time scalings, and special conformal transformations [26]. The conformal group is the largest symmetry group of Maxwell equations in free space. Conformally invariant results have hence the maximum possible validity in electromagnetism.

An important distinction is in order at this point. The use of the conformally invariant inner product ensures the maximal validity for average quantities as defined by Equation (1): The projection of the vector $|\mathcal{F}\rangle$ onto the vector $\Gamma|\mathcal{F}\rangle$ is equal to the projection of $|\tilde{\mathcal{F}}\rangle = T|\mathcal{F}\rangle$ onto $|\widetilde{\Gamma\mathcal{F}}\rangle = T\Gamma|\mathcal{F}\rangle$, i.e., $\langle\mathcal{F}|T^\dagger TT|\mathcal{F}\rangle = \langle\mathcal{F}|\Gamma|\mathcal{F}\rangle$, for any transformation T in the conformal group, where $T^\dagger$ is the Hermitian adjoint of T. Satisfying this demand amounts to showing that an inner product exists with respect to which the conformal group acts unitarily ($TT^\dagger = T^\dagger T = I$ for all T, where I is the identity) on the vector space of solutions of Maxwell equations [26]. Loosely speaking, this means that the value of the averages in Equation (1) will not change regardless of "the conformal point of view" or "conformal change of coordinate system". This will hold for average helicity, and also for average momentum, average angular momentum, etc.... A *different* matter is whether the average quantity in a conformally transformed field is the same as the average quantity in the initial field, for all conformal transformations. In this case, we are asking whether $|\mathcal{F}\rangle$ and $T|\mathcal{F}\rangle$ have the same average value of a given property Γ, i.e., whether

$$\langle\mathcal{F}|T^\dagger \Gamma T|\mathcal{F}\rangle = \langle\mathcal{F}|\Gamma|\mathcal{F}\rangle \text{ for all } T, \tag{2}$$

which is often not the case, such as for example when a Lorentz boost simultaneously changes the energy and momentum of a given field. Incidentally, it will be clear later that Equation (2) is actually met in the case of average helicity.

Writing down an explicit expression for Equation (1) requires us to choose an explicit representation for the vectors in $\mathbb{M}$ and the operators acting on them. We choose the following representation for the vectors in $\mathbb{M}$:

$$|\mathcal{F}\rangle \equiv \mathcal{F}(\mathbf{k}) = \begin{bmatrix} \mathbf{F}_+(\mathbf{k}) \\ \mathbf{F}_-(\mathbf{k}) \end{bmatrix}, \tag{3}$$

where the $\mathbf{F}_\pm(\mathbf{k})$ define the plane-wave components of a version of the Riemann-Silberstein vectors [5]

$$\frac{\mathbf{D}(\mathbf{r},t)}{\sqrt{2\epsilon_0}} \pm i\frac{\mathbf{B}(\mathbf{r},t)}{\sqrt{2\mu_0}} = \sqrt{\frac{\epsilon_0}{2}}\left[\mathbf{E}(\mathbf{r},t) \pm iZ_0\mathbf{H}(\mathbf{r},t)\right] = \mathbf{F}_\pm(\mathbf{r},t)$$

$$= \int_{\mathbb{R}^3-\{0\}} \frac{d\mathbf{k}}{\sqrt{(2\pi)^3}} \mathbf{F}_\pm(\mathbf{k})\exp(i\mathbf{k}\cdot\mathbf{r}-i\omega t), \text{ with } c_0\sqrt{\mathbf{k}\cdot\mathbf{k}} = c_0 k = \omega > 0, \tag{4}$$

where ϵ_0, μ_0, c_0, and $Z_0 = \sqrt{\mu_0/\epsilon_0}$ are the vacuum's permittivity, permeability, speed of light, and impedance, respectively, $\mathbf{k}$ is the wavevector, and $\omega = c_0 k = c_0\sqrt{\mathbf{k}\cdot\mathbf{k}}$ is the angular frequency. The $\mathbf{F}_{\pm}(\mathbf{k})$ can be further decomposed as $\mathbf{F}_{\pm}(\mathbf{k}) = \hat{\mathbf{e}}_{\pm}(\hat{\mathbf{k}})f_{\pm}(\mathbf{k})$, where $f_{\pm}(\mathbf{k})$ are complex-valued scalar functions and $\hat{\mathbf{e}}_{\pm}(\hat{\mathbf{k}})$ are the $\hat{\mathbf{k}}$-dependent polarization vectors for each handedness(helicity) (The $\hat{\mathbf{e}}_{\pm}(\hat{\mathbf{k}})$ can be obtained by the rotation of $(\pm\hat{\mathbf{x}} - i\hat{\mathbf{y}})/\sqrt{2}$, the two vectors corresponding to $\hat{\mathbf{k}} = \hat{\mathbf{z}}$: $\sqrt{2}\hat{\mathbf{e}}_{\pm}(\hat{\mathbf{k}}) = R_z(\phi)R_y(\theta)(\mp\hat{\mathbf{x}} - i\hat{\mathbf{y}})$, where $\theta = \arccos(k_z/|\mathbf{k}|)$ and $\phi = \arctan(k_y, k_x)$). We note that $\hat{\mathbf{k}} \cdot \hat{\mathbf{e}}_{\pm}(\hat{\mathbf{k}}) = 0$, which makes the $\mathbf{F}_{\pm}(\mathbf{k})[\mathbf{F}_{\pm}(\mathbf{r},t)]$ transverse functions, namely $\hat{\mathbf{k}} \cdot \mathbf{F}_{\pm}(\mathbf{k}) = \nabla \cdot \mathbf{F}_{\pm}(\mathbf{r},t) = 0$. The origin $\mathbf{k} = 0$ is removed in the integral in Equation (4) because we are considering electrodynamics and excluding electro- and magneto-statics, whereby $|\mathbf{k}| = \omega/c_0 = 0$ needs to be excluded. It important to note that *only positive frequencies are included* in Equation (4). This amounts to considering positive energies only, which is possible in electromagnetism since the photon is its own anti-particle. Only one sign of the energy (frequency) is needed because the same information is contained on both sides of the spectrum ([5] § 3.1 and [4]). When only positive frequencies are included, $\mathbf{D}(\mathbf{r},t)$, $\mathbf{B}(\mathbf{r},t)$, $\mathbf{E}(\mathbf{r},t)$ and $\mathbf{H}(\mathbf{r},t)$ in Equation (4) are complex-valued fields. With $\mathbf{X}$ standing for $\mathbf{D}$, $\mathbf{B}$, $\mathbf{E}$ or $\mathbf{H}$:

$$\mathbf{X}(\mathbf{r},t) = \int_{>0}^{\infty} \frac{d\omega}{\sqrt{2\pi}} \mathbf{X}^{\omega}(\mathbf{r}) \exp(-i\omega t). \tag{5}$$

We define the complex-valued fields so that the typical real-valued versions are obtained as

$$\mathcal{X}(\mathbf{r},t) = \int_{>0}^{\infty} \frac{d\omega}{\sqrt{2\pi}} \mathbf{X}^{\omega}(\mathbf{r}) \exp(-i\omega t) + \mathbf{X}^{\omega}(\mathbf{r})^* \exp(i\omega t). \tag{6}$$

The restriction to positive frequencies is particularly consequential for the treatment of helicity, the generalized polarization handednesses of the field. One of the advantages of the Riemann-Silberstein vectors is their ability to encode the helicity content of the field. They are the eigenstates of the helicity operator and potentially allow for the splitting of the two polarization handednesses in any field, including near and evanescent fields. However, when they are defined by means of real-valued fields, as in $\frac{\mathcal{D}(\mathbf{r},t)}{\sqrt{2\epsilon_0}} \pm i\frac{\mathcal{B}(\mathbf{r},t)}{\sqrt{2\mu_0}}$, their use for splitting the two helicities is not as simple as it becomes when complex-valued fields are used. With real-valued fields we have that the two $\pm$ fields determine each other through complex conjugation $\left[\frac{\mathcal{D}(\mathbf{r},t)}{\sqrt{2\epsilon_0}} + i\frac{\mathcal{B}(\mathbf{r},t)}{\sqrt{2\mu_0}}\right]^* = \frac{\mathcal{D}(\mathbf{r},t)}{\sqrt{2\epsilon_0}} - i\frac{\mathcal{B}(\mathbf{r},t)}{\sqrt{2\mu_0}}$, which is at odds with the a priori physical independence of the two helicity components of the electromagnetic field. For example, the complex conjugation connection means that the two $\pm$ squared norms $\left|\frac{\mathcal{D}(\mathbf{r},t)}{\sqrt{2\epsilon_0}} \pm i\frac{\mathcal{B}(\mathbf{r},t)}{\sqrt{2\mu_0}}\right|^2$, which could intuitively be thought of as the $(\mathbf{r},t)$-local helicity intensities, become equal at all space–time points $\left|\frac{\mathcal{D}(\mathbf{r},t)}{\sqrt{2\epsilon_0}} + i\frac{\mathcal{B}(\mathbf{r},t)}{\sqrt{2\mu_0}}\right|^2 = \left|\frac{\mathcal{D}(\mathbf{r},t)}{\sqrt{2\epsilon_0}} - i\frac{\mathcal{B}(\mathbf{r},t)}{\sqrt{2\mu_0}}\right|^2$. This contradicts, for example, the fact that there can be electromagnetic fields containing only one of the two helicities, e.g., any linear combination of plane-waves with the same polarization handedness. The restriction to positive frequencies overcomes these limitations: In Equation (4) $\mathbf{F}_{+}(\mathbf{r},t)$ contains no information about $\mathbf{F}_{-}(\mathbf{r},t)$, in particular $\mathbf{F}_{+}(\mathbf{r},t)^* \neq \mathbf{F}_{-}(\mathbf{r},t)$.

The choice of the representation in Equation (3), where the two helicity components are distinguished, as opposed to other more common possibilities where the electric and magnetic fields are distinguished, can also be motivated by the transformation properties of the two different options with respect to the conformal group. Namely helicity is invariant under the conformal group [27,28], i.e., all the generators of the conformal group commute with the helicity operator. This was established by Mack and Todorov in Ref. [27] when they showed that a Casimir operator of the conformal group is linearly related to the helicity operator. The invariance can also be inferred from the facts that (i) helicity is ultimately proportional to the cosine of an angle (The cosine of the angle between the vector of spin-1 matrices $\mathbf{S}$ and the linear momentum operator $\mathbf{P}$ is defined as $\cos(\angle(\mathbf{S},\mathbf{P})) = \frac{\mathbf{S}\cdot\mathbf{P}}{|\mathbf{P}||\mathbf{S}|}$. Then, using the definition of the helicity operator Λ in Equation (8) we can write $\Lambda = \cos(\angle(\mathbf{S},\mathbf{P}))|\mathbf{S}|$.

However, the action of $|\mathbf{S}|$ on members of $\mathbb{M}$ is trivial since $|\mathbf{S}|^2 = (S_1^2 + S_2^2 + S_3^2) = 2I$ where I is the identity. This can be seen in ([29] Equation (5.54).), and is readily verified by direct calculation using the spin-1 matrices.), and that, (ii) the preservation of angles is guaranteed by conformal transformations. In the representation of Equation (3) this invariance means that no matter which conformal transformation is applied to $|\mathcal{F}\rangle$, the $\mathbf{F}_+(\mathbf{k})$ upper components of $\mathcal{F}(\mathbf{k})$ will never end up on the lower part, and vice-versa. This reduces the algebraic complexity of some expressions and manipulations. In sharp contrast to this, what is meant by electric and magnetic fields is not conformally invariant. Actually, the meanings of "electric" and "magnetic" are not even relativistically invariant since electric and magnetic fields are intermixed by Lorentz boosts ([30] Equation (11.149)). An important physical fact about helicity can be deduced from its conformal invariance. Since the helicity operator (Λ) commutes with any T in the conformal group ($T\Lambda = \Lambda T$), and T is unitary with respect to the chosen inner product ($TT^\dagger = T^\dagger T = I$), we can readily see that Equation (2) is met ($\langle \mathcal{F}|T^\dagger \Lambda T|\mathcal{F}\rangle \stackrel{\Lambda T = T\Lambda}{=} \langle \mathcal{F}|T^\dagger T \Lambda|\mathcal{F}\rangle \stackrel{T^\dagger T = I}{=} \langle \mathcal{F}|\Gamma|\mathcal{F}\rangle$.): The average helicity of a conformally transformed field is the same as the average helicity of the initial field.

Let us go on to computing the average helicity of a given field as a conformally invariant inner product. We will explicitly keep the constants ϵ_0, μ_0, c_0, and Z_0 in the expressions, and use the four fields $\mathbf{D}$, $\mathbf{B}$, $\mathbf{E}$ and $\mathbf{H}$. These choices [21,23] facilitate the re-use of the formulas when a description such as the one in Equation (4) is possible in a non-vacuum background, such as for example in an infinite isotropic and homogeneous linear medium.

Following Gross [26], and Bialynicki-Birula ([31] § 9) and ([5] § 5), the definitions in Equations (1) and (3) allow us to write the average value of any property Γ as (Note: This is seen by comparing Equation (4) with ([5] Equations (4.11)–(4.12)), and Equation (7) particularized to the energy operator $\Gamma \to H = \begin{bmatrix} \omega I_{3\times 3} & 0_{3\times 3} \\ 0_{3\times 3} & \omega I_{3\times 3} \end{bmatrix}$ with ([5] Equation (4.13)), and setting $\hbar = 1$.)

$$\langle \mathcal{F}|\Gamma|\mathcal{F}\rangle = \int_{\mathbb{R}^3 - \{0\}} \frac{d\mathbf{k}}{c_0|\mathbf{k}|} \mathcal{F}(\mathbf{k})^\dagger \Gamma \mathcal{F}(\mathbf{k}), \tag{7}$$

where $\dagger$ means transpose conjugate. Equation (7) is an explicit expression of the conformally invariant inner product between $|\mathcal{F}\rangle$ and $\Gamma|\mathcal{F}\rangle$ (Note: When Equation (7) is brought to the $(\mathbf{r}, t)$ domain, the $\mathbf{k}$-local expression results in the double integral $\int_{\mathbb{R}^3} d\mathbf{r} \int_{\mathbb{R}^3} d\bar{\mathbf{r}}$ of a manifestly non-$\mathbf{r}$-local integrand including a term such as $1/|\mathbf{r} - \bar{\mathbf{r}}|^2$ (see [26] Equation (6) and [5] Equation (5.7))).

We are now ready to focus our attention on the average value of helicity. The helicity operator Λ is defined as the projection of the angular momentum operator vector $\mathbf{J}$ onto the direction of the linear momentum operator vector $\mathbf{P}$:

$$\Lambda = \frac{\mathbf{J} \cdot \mathbf{P}}{|\mathbf{P}|} = \frac{\mathbf{S} \cdot \mathbf{P}}{|\mathbf{P}|}, \tag{8}$$

where for electromagnetism, $\mathbf{S}$ is the vector of spin-1 matrices (The second equality can be seen to follow, for example, from considering the coordinate representation of the angular momentum and linear momentum operator vectors, (ref. [5] Equations (5.24) and (5.25)): $\mathbf{J} \equiv -i\mathbf{r} \times \nabla + \mathbf{S}$, $\mathbf{P} \equiv -i\nabla$. Their inner product then reads $\mathbf{J} \cdot \mathbf{P} \equiv -(\mathbf{r} \times \nabla) \cdot \nabla - i\mathbf{S} \cdot \nabla$. The first term vanishes since it is the divergence of a curl).

We start by particularizing Equation (7) to the helicity operator Λ.

$$\langle \mathcal{F}|\Lambda|\mathcal{F}\rangle = \int_{\mathbb{R}^3 - \{0\}} \frac{d\mathbf{k}}{c_0|\mathbf{k}|} \mathcal{F}(\mathbf{k})^\dagger \Lambda \mathcal{F}(\mathbf{k}) = \int_{\mathbb{R}^3 - \{0\}} \frac{d\mathbf{k}}{c_0|\mathbf{k}|} \begin{bmatrix} \mathbf{F}_+(\mathbf{k}) \\ \mathbf{F}_-(\mathbf{k}) \end{bmatrix}^\dagger \begin{bmatrix} i\hat{\mathbf{k}} \times & 0 \\ 0 & i\hat{\mathbf{k}} \times \end{bmatrix} \begin{bmatrix} \mathbf{F}_+(\mathbf{k}) \\ \mathbf{F}_-(\mathbf{k}) \end{bmatrix}, \tag{9}$$

where the last expression contains the explicit form of the helicity operator in our choice of representation (This follows from the definition of helicity in Equation (8): $\mathbf{S} \cdot \mathbf{P}/|\mathbf{P}| \equiv i\hat{\mathbf{k}} \times$, where the equivalence follows from applying ([5] Equation (2.2))) in momentum space where

$\mathbf{P} \to \mathbf{k} \implies \mathbf{P}/|\mathbf{P}| \to \hat{\mathbf{k}}$). We now use the fact that the $\mathbf{F}_\pm(\mathbf{k})$ are eigenstates of helicity, namely $i\hat{\mathbf{k}} \times \mathbf{F}_\pm(\mathbf{k}) = \pm \mathbf{F}_\pm(\mathbf{k})$, to write

$$\langle \mathcal{F}|\Lambda|\mathcal{F}\rangle = \int_{\mathbb{R}^3 - \{0\}} \frac{d\mathbf{k}}{c_0|\mathbf{k}|} \mathbf{F}_+(\mathbf{k})^\dagger \mathbf{F}_+(\mathbf{k}) - \mathbf{F}_-(\mathbf{k})^\dagger \mathbf{F}_-(\mathbf{k}) = \int_{\mathbb{R}^3 - \{0\}} \frac{d\mathbf{k}}{c_0|\mathbf{k}|} |\mathbf{F}_+(\mathbf{k})|^2 - |\mathbf{F}_-(\mathbf{k})|^2. \quad (10)$$

We will now show that Equation (10) is equivalent to the most common integral expression of the helicity average. To such end, and taking advantage of the fact that $k \neq 0$, we define the helicity potentials

$$\mathcal{V}(\mathbf{k}) = \frac{1}{ikc_0}\mathcal{F}(\mathbf{k}) = \frac{1}{ikc_0}\begin{bmatrix} \mathbf{F}_+(\mathbf{k}) \\ \mathbf{F}_-(\mathbf{k}) \end{bmatrix} = \begin{bmatrix} \mathbf{V}_+(\mathbf{k}) \\ \mathbf{V}_-(\mathbf{k}) \end{bmatrix}, \quad (11)$$

which in the $(\mathbf{r},t)$ representation, and recalling that $-i\omega \to \partial_t$, are seen to be related to $\mathbf{F}_\pm(\mathbf{r},t)$ as

$$-\partial_t \mathbf{V}_\pm(\mathbf{r},t) = \mathbf{F}_\pm(\mathbf{r},t) = \sqrt{\frac{\epsilon_0}{2}}\left[\mathbf{E}(\mathbf{r},t) \pm iZ_0\mathbf{H}(\mathbf{r},t)\right], \quad (12)$$

from where we can use ([21] Equation (2)), namely $-\partial_t \mathbf{C}(\mathbf{r},t) = \mathbf{H}(\mathbf{r},t)$ and $-\partial_t \mathbf{A}(\mathbf{r},t) = \mathbf{E}(\mathbf{r},t)$, to recognize that these helicity potentials are linear combinations of complex versions of the transverse real-valued "magnetic" $\mathbf{A}(\mathbf{r},t)$ and "electric" $\mathbf{C}(\mathbf{r},t)$ potentials [2,8,11–13,17,25].

$$\mathbf{V}_\pm(\mathbf{r},t) = \sqrt{\frac{\epsilon_0}{2}}\left[\mathbf{A}(\mathbf{r},t) \pm iZ_0\mathbf{C}(\mathbf{r},t)\right]. \quad (13)$$

Appendix A contains some background information about the electric potential. Linear combinations very similar to Equation (13) have been recently introduced by Elbistan et al. in Ref. [17], albeit using real-valued vector functions instead of our complex $\mathbf{A}(\mathbf{r},t)$ and $\mathbf{C}(\mathbf{r},t)$. As previously discussed, this difference is relevant for treating helicity. When real-valued fields are used in the right hand side of Equation (13), it follows that $\mathbf{V}_+(\mathbf{r},t)^* = \mathbf{V}_-(\mathbf{r},t)$, which ultimately leads to a zero value of the average helicity as reported in [17].

It is also worth pointing out that the $\mathbf{V}_\pm(\mathbf{k})$ functions are transverse, i.e., $\hat{\mathbf{k}} \cdot \mathbf{V}_\pm(\mathbf{k}) = 0$, which follows from Equation (11) and the previously mentioned property $\hat{\mathbf{k}} \cdot \mathbf{F}_\pm(\mathbf{k})$. The helicity potentials only contain the transverse degrees of freedom, the same as the free electromagnetic field, which ensures that the results obtained using $\mathbf{V}_\pm(\mathbf{k})$ are gauge independent.

We proceed by using Equation (11) and the central expression in Equation (10) to obtain

$$\langle \mathcal{F}|\Lambda|\mathcal{F}\rangle = i\int_{\mathbb{R}^3 - \{0\}} d\mathbf{k}\, \mathbf{F}_+(\mathbf{k})^\dagger \mathbf{V}_+(\mathbf{k}) - \mathbf{F}_-(\mathbf{k})^\dagger \mathbf{V}_-(\mathbf{k}). \quad (14)$$

Equation (14) can now be brought to the $(\mathbf{r},t)$ domain as follows. First, we apply the substitutions $\mathbf{F}_\pm(\mathbf{k}) \to \mathbf{F}_\pm(\mathbf{k})\exp(-ic_0k)$, and $\mathbf{V}_\pm(\mathbf{k}) \to \mathbf{V}_\pm(\mathbf{k})\exp(-ic_0k)$

$$\langle \mathcal{F}|\Lambda|\mathcal{F}\rangle = $$
$$i\int_{\mathbb{R}^3 - \{0\}} d\mathbf{k}\, \left[\mathbf{F}_+(\mathbf{k})\exp(-ic_0k)\right]^\dagger \left[\mathbf{V}_+(\mathbf{k})\exp(-ic_0k)\right] - \left[\mathbf{F}_-(\mathbf{k})\exp(-ic_0k)\right]^\dagger \left[\mathbf{V}_-(\mathbf{k})\exp(-ic_0k)\right]. \quad (15)$$

These changes do not affect the result, but allow us to see from Equation (4) that the $\mathbf{F}_\pm(\mathbf{k})\exp(-ic_0k)$ are the three-dimensional Fourier transforms ($\mathbf{r} \to \mathbf{k}$) of $\mathbf{F}_\pm(\mathbf{r},t)$. The same relation holds between $\mathbf{V}_\pm(\mathbf{k})\exp(-ic_0k)$ and $\mathbf{V}_\pm(\mathbf{r},t)$. We can now apply applying Parseval's theorem, i.e., the unitarity of the inverse Fourier transform $\mathbf{k} \to \mathbf{r}$, to each of the two terms in the subtraction in Equation (15):

$$\langle \mathcal{F}|\Lambda|\mathcal{F}\rangle = i\int_{\mathbb{R}^3} d\mathbf{r}\, \mathbf{F}_+(\mathbf{r},t)^\dagger \mathbf{V}_+(\mathbf{r},t) - \mathbf{F}_-(\mathbf{r},t)^\dagger \mathbf{V}_-(\mathbf{r},t), \quad (16)$$

where the integrand is local in $\mathbf{r}$. We show in Appendix A that when Equation (10) is brought to the $(\mathbf{r},t)$ domain instead, the $1/|\mathbf{k}|$ term results in the double integral $\int_{\mathbb{R}^3} d\mathbf{r} \int_{\mathbb{R}^3} d\bar{\mathbf{r}}$ of a manifestly non-r-local integrand including a term such as $1/|\mathbf{r} - \bar{\mathbf{r}}|^2$. The inconvenient $1/|\mathbf{k}|$ term is absorbed in the definition of the potentials in Equation (11).

To further approach the most common expression of the average helicity, we now substitute

$$\mathbf{F}_\pm(\mathbf{r},t) = \frac{\mathbf{D}(\mathbf{r},t)}{\sqrt{2\epsilon_0}} \pm i\frac{\mathbf{B}(\mathbf{r},t)}{\sqrt{2\mu_0}}, \quad \mathbf{V}_\pm(\mathbf{r},t) = \sqrt{\frac{\epsilon_0}{2}}\left[\mathbf{A}(\mathbf{r},t) \pm iZ_0\mathbf{C}(\mathbf{r},t)\right], \tag{17}$$

into Equation (16) and obtain

$$\begin{aligned}\langle \mathcal{F}|\Lambda|\mathcal{F}\rangle &= \frac{i}{2}\int_{\mathbb{R}^3} d\mathbf{r} \\ & \left[\mathbf{D}(\mathbf{r},t)^\dagger \mathbf{A}(\mathbf{r},t) + iZ_0\mathbf{D}(\mathbf{r},t)^\dagger \mathbf{C}(\mathbf{r},t) - \frac{i}{Z_0}\mathbf{B}(\mathbf{r},t)^\dagger \mathbf{A}(\mathbf{r},t) + \mathbf{B}(\mathbf{r},t)^\dagger \mathbf{C}(\mathbf{r},t)\right] \\ &- \left[\mathbf{D}(\mathbf{r},t)^\dagger \mathbf{A}(\mathbf{r},t) - iZ_0\mathbf{D}(\mathbf{r},t)^\dagger \mathbf{C}(\mathbf{r},t) + \frac{i}{Z_0}\mathbf{B}(\mathbf{r},t)^\dagger \mathbf{A}(\mathbf{r},t) + \mathbf{B}(\mathbf{r},t)^\dagger \mathbf{C}(\mathbf{r},t)\right] \\ &= \int_{\mathbb{R}^3} d\mathbf{r}\, \frac{1}{Z_0}\mathbf{B}(\mathbf{r},t)^\dagger \mathbf{A}(\mathbf{r},t) - Z_0\mathbf{D}(\mathbf{r},t)^\dagger \mathbf{C}(\mathbf{r},t),\end{aligned} \tag{18}$$

which is a complex version of the well-known integral for the average helicity featuring real-valued fields, as found e.g., in ([21] Equation (6)). Appendix B shows that the results of the complex and real versions coincide.

The **k**-domain expressions in Equations (10) and (14), and the $(\mathbf{r},t)$-domain expression in Equation (16) produce the correct result. We now obtain $(\mathbf{r},\omega)$-domain expressions. The time-harmonic decomposition is often used in both theoretical investigations and numerical computations.

We start by noting that the result of the integral in Equation (16) is independent of time (Indeed, the simplifying arbitrary choice $t = 0$ is made by Gross in [26] for evaluating the inner product with integrals featuring $(\mathbf{r},t)$-dependent integrands). The time independence of $\langle\mathcal{F}|\Lambda|\mathcal{F}\rangle$, manifest in Equations (10) and (14), is ultimately due the fact that $\langle\mathcal{F}|\Lambda|\mathcal{F}\rangle$ must be invariant under time translations since such transformations are in the conformal group, i.e., $\langle\mathcal{F}|\Lambda|\mathcal{F}\rangle$ cannot depend on time. This can be exploited to obtain expressions for $\langle\mathcal{F}|\Lambda|\mathcal{F}\rangle$ involving the time-harmonic decomposition of the fields. To such end, we go back to Equation (16), and expand each term in the integrand into their frequency components

$$\begin{aligned}\langle \mathcal{F}|\Lambda|\mathcal{F}\rangle &= i\int_{\mathbb{R}^3} d\mathbf{r} \\ & \left[\int_{>0}^\infty \frac{d\omega}{\sqrt{2\pi}} \mathbf{F}_+^\omega(\mathbf{r})\exp(-i\omega t)\right]^\dagger \left[\int_{>0}^\infty \frac{d\omega}{\sqrt{2\pi}} \mathbf{V}_+^\omega(\mathbf{r})\exp(-i\omega t)\right] \\ &- \left[\int_{>0}^\infty \frac{d\omega}{\sqrt{2\pi}} \mathbf{F}_-^\omega(\mathbf{r})\exp(-i\omega t)\right]^\dagger \left[\int_{>0}^\infty \frac{d\omega}{\sqrt{2\pi}} \mathbf{V}_-^\omega(\mathbf{r})\exp(-i\omega t)\right] \\ &= i\int_{\mathbb{R}^3} d\mathbf{r} \int_{>0}^\infty \frac{d\omega}{\sqrt{2\pi}} \int_{>0}^\infty \frac{d\bar{\omega}}{\sqrt{2\pi}} \left[\mathbf{F}_+^{\bar{\omega}}(\mathbf{r})^\dagger \mathbf{V}_+^\omega(\mathbf{r}) - \mathbf{F}_-^{\bar{\omega}}(\mathbf{r})^\dagger \mathbf{V}_-^\omega(\mathbf{r})\right] \exp(-i(\omega-\bar{\omega})t).\end{aligned} \tag{19}$$

Let us examine the last line of Equation (19). Because $\langle\mathcal{F}|\Lambda|\mathcal{F}\rangle$ cannot depend on time, and since the two helicities are independent of each other, it follows that only the $\omega = \bar{\omega}$ components can contribute to the end result. This allows us to obtain the following three equivalent expressions:

$$\begin{aligned}(2\pi)\langle\mathcal{F}|\Lambda|\mathcal{F}\rangle &= i\int_{\mathbb{R}^3} d\mathbf{r} \int_{>0}^\infty d\omega\, \mathbf{F}_+^\omega(\mathbf{r})^\dagger \mathbf{V}_+^\omega(\mathbf{r}) - \mathbf{F}_-^\omega(\mathbf{r})^\dagger \mathbf{V}_-^\omega(\mathbf{r}) \\ &= \int_{\mathbb{R}^3} d\mathbf{r} \int_{>0}^\infty d\omega\, \frac{1}{\omega}\left(|\mathbf{F}_+^\omega(\mathbf{r})|^2 - |\mathbf{F}_-^\omega(\mathbf{r})|^2\right) \\ &= \int_{\mathbb{R}^3} d\mathbf{r} \int_{>0}^\infty d\omega\, \omega\left(|\mathbf{V}_+^\omega(\mathbf{r})|^2 - |\mathbf{V}_-^\omega(\mathbf{r})|^2\right),\end{aligned} \tag{20}$$

where the equalities readily follow from $\mathbf{F}_\pm^\omega(\mathbf{r}) = i\omega \mathbf{V}_\pm^\omega(\mathbf{r})$, which follows from Equation (12).

Expressions that are local in $\mathbf{r}$, such as Equations (14) and (20), justify the consideration of the average helicity in a finite volume $\mathbb{D}$. This then allows use of the corresponding expressions in practical situations where numerical solvers calculate the fields in finite regions of space. The expressions in Equation (20) are particularly adapted to the output of finite-element-method solvers such as COMSOL and JCM, which use the time-harmonic decomposition of $\mathbf{r}$-dependent fields.

Finally, regarding applications, the electromagnetic helicity is particularly relevant in chiral light–matter interactions. Among these, the interaction of the field with chiral molecules is one of the most researched cases, partly because the optical sensing of chiral molecules is important in chemistry and pharmaceutical applications. In Appendix C we use the above formalism to derive expressions for computing the circular dichroism signal for two different settings of the light-molecule interaction: The 6 × 6 dipolarizability tensor and the T-matrix.

In conclusion, several equivalent expressions for the average value of the electromagnetic helicity of a given field have been obtained from a starting point featuring maximal electromagnetic invariance, i.e., from an expression whose result is invariant under the conformal group. Some of the obtained expressions can be conveniently evaluated using the outputs of common Maxwell solvers.

Funding: Partially funded by the Deutsche Forschungsgemeinschaft (DFG, German Research Foundation)—Project-ID 258734477—SFB 1173. I would also like to acknowledge support by KIT through the Virtual Materials Design (VIRTMAT) project by the Helmholtz Association via the Helmholtz program Science and Technology of Nanosystems (STN).

Conflicts of Interest: The author declares no conflict of interests.

Appendix A. The Use of Two Potentials for Obtaining r-Local Integrands

The use of two potentials, one magnetic and one electric, has a long tradition in the studies of helicity and of the symmetry generated by the helicity operator: Electromagnetic duality [2,8,11–13,17,25]. Duality can be seen as the underlying reason for adding an electric potential next to the magnetic one.

In the absence of sources, the (real-valued) electric potential $\mathcal{C}(\mathbf{r},t)$ is typically defined by first fixing its transverse part

$$\mathcal{E}_\perp(\mathbf{r},t) = -\nabla \times \mathcal{C}(\mathbf{r},t), \tag{A1}$$

where $\mathcal{E}_\perp(\mathbf{r},t)$ is the transverse electric field, and then exploiting the fact that $\mathcal{C}(\mathbf{r},t)$ has its own gauge freedom [2,13,22] to fix the longitudinal part by a choice of gauge. When the radiation gauge ($\nabla \cdot \mathcal{C}(\mathbf{r},t) = 0$) is chosen, $\mathcal{C}(\mathbf{r},t)$ becomes a transverse field, containing the same kind of degrees of freedom as the radiation electromagnetic fields. The electric potential has also been used in the presence of sources [13,22,25]. The choice of the radiation gauge is also adequate in this case, since it can be shown that the longitudinal degrees of freedom of the field can always be adscribed to the sources instead ([32] I.B.5) and ([33] Chap. XXI, § 22).

In the particular context of integral expressions for the average electromagnetic helicity, the introduction of potentials allows to obtain $\mathbf{r}$-local integrands. This has been shown in the main text in the derivations leading to the $\mathbf{r}$-local Equation (16). We will know bring Equation (10), which does not involve the potentials, to the $(\mathbf{r},t)$ domain and see how precisely the non-$\mathbf{r}$-locality arises. We start hence from Equation (10), which only contains the $\mathbf{F}_\pm(\mathbf{k})$ fields, and apply the non-result-altering substitutions $\mathbf{F}_\pm(\mathbf{k}) \to \mathbf{F}_\pm(\mathbf{k}) \exp(-ic_0 k)$:

$$\langle \mathcal{F}|\Lambda|\mathcal{F}\rangle = \int_{\mathbb{R}^3-\{0\}} \frac{d\mathbf{k}}{c_0|\mathbf{k}|} \left[[\mathbf{F}_+(\mathbf{k})\exp(-ic_0 k)]^\dagger [\mathbf{F}_+(\mathbf{k})\exp(-ic_0 k)] - [\mathbf{F}_-(\mathbf{k})\exp(-ic_0 k)]^\dagger [\mathbf{F}_-(\mathbf{k})\exp(-ic_0 k)] \right]. \tag{A2}$$

We will first focus on the first term of the integrand, which we consider as the $\mathbf{k}$-point-wise inner product of two functions: $\mathbf{F}_+(\mathbf{k})\exp(-ic_0 k)$ and $\mathbf{F}_+(\mathbf{k})\exp(-ic_0 k)/|\mathbf{k}|$. In order to apply Parseval's theorem and bring Equation (A2) to the $(\mathbf{r},t)$ domain, the inverse 3D Fourier transforms of the two functions are needed. We know from the definitions in Equation (4) that the inverse 3D Fourier transform of $\mathbf{F}_+(\mathbf{k})\exp(-ic_0k)$ is $\mathbf{F}_+(\mathbf{r},t)$. The inverse transform of the product $\mathbf{F}_+(\mathbf{k})\exp(-ic_0k) \times \frac{1}{|\mathbf{k}|}$ can be obtained using the convolution theorem and the inverse Fourier transform, denoted by $\mathscr{F}_{3D}^{-1}\{\cdot\}$, of each of the two factors (see Equations (B.3) and (B.4) and Tab. II in ([32] I.B.2)):

$$\mathscr{F}_{3D}^{-1}\left\{\mathbf{F}_+(\mathbf{k})\exp(-ic_0|\mathbf{k}|) \times \frac{1}{|\mathbf{k}|}\right\} = \frac{1}{2\pi^2}\int_{\mathbb{R}^3} d\bar{\mathbf{r}}\, \mathbf{F}_+(\bar{\mathbf{r}},t) \times \frac{1}{|\mathbf{r}-\bar{\mathbf{r}}|^2}. \tag{A3}$$

Using Equation (A3) and its obvious counterpart for the second term in the integrand of Equation (A2), we can use Parseval's theorem to write:

$$\langle\mathcal{F}|\Lambda|\mathcal{F}\rangle = \frac{1}{2\pi^2 c_0}\int_{\mathbb{R}^3} d\mathbf{r} \int_{\mathbb{R}^3} d\bar{\mathbf{r}}\, \frac{\mathbf{F}_+(\mathbf{r},t)^\dagger \mathbf{F}_+(\bar{\mathbf{r}},t) - \mathbf{F}_-(\mathbf{r},t)^\dagger \mathbf{F}_-(\bar{\mathbf{r}},t)}{|\mathbf{r}-\bar{\mathbf{r}}|^2}. \tag{A4}$$

As explained in the main text, the typically undesired non-locality of the integrand in Equation (A4) is avoided by the introduction of the helicity potentials, since they absorb the $1/|\mathbf{k}|$ term into their definition. We note that previously existing non-local expressions for average electromagnetic helicity, like ([34] Equation (65)) and ([24] Equation (36)), can be shown to be equivalent to Equation (A4). The same arguments show why the introduction of the magnetic and electric potentials, $\mathcal{A}(\mathbf{r},t)$ and $\mathcal{C}(\mathbf{r},t)$, results in a r-local integrand in the typical definition of average electromagnetic helicity. Finally, we note that mixed formulations exist where only one of the two potentials is used, which still result in non-r-local integrands ([3] Equation (2.6)).

Appendix B. Equivalence between Complex and Real Versions

In this Appendix we show that Equation (18) of the main text, featuring complex-valued fields

$$\langle\mathcal{F}|\Lambda|\mathcal{F}\rangle = \int_{\mathbb{R}^3} d\mathbf{r}\, \frac{1}{Z_0}\mathbf{B}(\mathbf{r},t)^\dagger \mathbf{A}(\mathbf{r},t) - Z_0 \mathbf{D}(\mathbf{r},t)^\dagger \mathbf{C}(\mathbf{r},t), \tag{A5}$$

featuring complex-valued fields is equivalent to the well-known integral for the average helicity featuring real-valued fields, as found e.g., in ([21] Equation (6)):

$$\frac{1}{2}\int_{\mathbb{R}^3} d\mathbf{r}\, \frac{1}{Z_0}\mathcal{B}(\mathbf{r},t)\cdot\mathcal{A}(\mathbf{r},t) - Z_0\mathcal{D}(\mathbf{r},t)\cdot\mathcal{C}(\mathbf{r},t). \tag{A6}$$

We will use properties of complex-valued vector fields whose Fourier transforms contain only positive frequencies, as defined in Equation (5) for $\mathbf{X}$ standing for $\mathbf{A}$, $\mathbf{B}$, $\mathbf{C}$, $\mathbf{D}$, and $\mathbf{E}$:

$$\mathbf{X}(\mathbf{r},t) = \int_{>0}^\infty \frac{d\omega}{\sqrt{2\pi}}\, \mathbf{X}^\omega(\mathbf{r})\exp(-i\omega t). \tag{A7}$$

The real and imaginary parts of $\mathbf{X}(\mathbf{r},t) = \mathbf{X}_{\mathrm{re}}(\mathbf{r},t) + i\mathbf{X}_{\mathrm{im}}(\mathbf{r},t)$ are related by the Hilbert transform, and then their Fourier transforms, denoted by $\mathscr{F}\{\cdot\}$, meet ([35] p. 49):

$$\mathscr{F}\{\mathbf{X}_{\mathrm{im}}(\mathbf{r},t)\} = (-i\,\mathrm{sign}\,\omega)\,\mathscr{F}\{\mathbf{X}_{\mathrm{re}}(\mathbf{r},t)\}. \tag{A8}$$

We now proceed by writing Equation (A5) using the real and imaginary parts of each field. Since the end result of the integral must be a real number, we can already discard the imaginary part of the integrand:

$$\langle \mathcal{F}|\Lambda|\mathcal{F}\rangle = \int_{\mathbb{R}^3} d\mathbf{r} \, \frac{1}{Z_0} \mathbf{B}_{\text{re}}(\mathbf{r},t)^\dagger \mathbf{A}_{\text{re}}(\mathbf{r},t) + \frac{1}{Z_0} \mathbf{B}_{\text{im}}(\mathbf{r},t)^\dagger \mathbf{A}_{\text{im}}(\mathbf{r},t)$$
$$- Z_0 \mathbf{D}_{\text{re}}(\mathbf{r},t)^\dagger \mathbf{C}_{\text{re}}(\mathbf{r},t) - Z_0 \mathbf{D}_{\text{im}}(\mathbf{r},t)^\dagger \mathbf{C}_{\text{im}}(\mathbf{r},t) =$$
$$\int_{\mathbb{R}^3} d\mathbf{r} \left[\frac{1}{Z_0} \mathbf{B}_{\text{re}}(\mathbf{r},t)^\dagger \mathbf{A}_{\text{re}}(\mathbf{r},t) - Z_0 \mathbf{D}_{\text{re}}(\mathbf{r},t)^\dagger \mathbf{C}_{\text{re}}(\mathbf{r},t) \right] + \left[\frac{1}{Z_0} \mathbf{B}_{\text{im}}(\mathbf{r},t)^\dagger \mathbf{A}_{\text{im}}(\mathbf{r},t) - Z_0 \mathbf{D}_{\text{im}}(\mathbf{r},t)^\dagger \mathbf{C}_{\text{im}}(\mathbf{r},t) \right]. \quad (A9)$$

We will now show that the two expressions in square brackets produce the same contribution. To such end, let us focus on one of their terms and use the time-harmonic decomposition (Note: The one sided integral in Equation (6) can be written as a two sided integral over the frequency axis and the familiar result $\mathbf{X}^\omega(\mathbf{r})^* = \mathbf{X}^{-\omega}(\mathbf{r})$ is recovered).

$$\int_{\mathbb{R}^3} d\mathbf{r} \, \mathbf{B}_{\text{im}}(\mathbf{r},t)^\dagger \mathbf{A}_{\text{im}}(\mathbf{r},t) =$$
$$\int_{\mathbb{R}^3} d\mathbf{r} \left[\int_{-\infty}^{\infty} \frac{d\omega}{\sqrt{2\pi}} \mathscr{F}\{\mathbf{B}_{\text{im}}(\mathbf{r},t)\} \exp(-i\omega t) \right]^\dagger \left[\int_{-\infty}^{\infty} \frac{d\omega}{\sqrt{2\pi}} \mathscr{F}\{\mathbf{A}_{\text{im}}(\mathbf{r},t)\} \exp(-i\omega t) \right]. \quad (A10)$$

The same considerations that take the last line of Equation (19) to the first line of Equation (20) can be used to write:

$$(2\pi) \int_{\mathbb{R}^3} d\mathbf{r} \, \mathbf{B}_{\text{im}}(\mathbf{r},t)^\dagger \mathbf{A}_{\text{im}}(\mathbf{r},t) = \int_{\mathbb{R}^3} d\mathbf{r} \int_{-\infty}^{\infty} d\omega \, \mathscr{F}\{\mathbf{B}_{\text{im}}(\mathbf{r},t)\}^\dagger \mathscr{F}\{\mathbf{A}_{\text{im}}(\mathbf{r},t)\}. \quad (A11)$$

We now use Equation (A8) on the last expression in Equation (A11)

$$(2\pi) \int_{\mathbb{R}^3} d\mathbf{r} \, \mathbf{B}_{\text{im}}(\mathbf{r},t)^\dagger \mathbf{A}_{\text{im}}(\mathbf{r},t) =$$
$$\int_{\mathbb{R}^3} d\mathbf{r} \int_{-\infty}^{\infty} d\omega \left[(-i\,\text{sign}\,\omega) \mathscr{F}\{\mathbf{B}_{\text{re}}(\mathbf{r},t)\} \right]^\dagger \left[(-i\,\text{sign}\,\omega) \mathscr{F}\{\mathbf{A}_{\text{re}}(\mathbf{r},t)\} \right] = \quad (A12)$$
$$\int_{\mathbb{R}^3} d\mathbf{r} \int_{-\infty}^{\infty} d\omega \, \mathscr{F}\{\mathbf{B}_{\text{re}}(\mathbf{r},t)\}^\dagger \mathscr{F}\{\mathbf{A}_{\text{re}}(\mathbf{r},t)\} = (2\pi) \int_{\mathbb{R}^3} d\mathbf{r} \, \mathbf{B}_{\text{re}}(\mathbf{r},t)^\dagger \mathbf{A}_{\text{re}}(\mathbf{r},t),$$

where the last equality follows by comparison with Equation (A11). Equation (A12) shows that the two expressions inside the square brackets in Equation (A9) produce the same contribution, since the steps leading to Equation (A12) can be applied to any of the product terms. We can hence write:

$$\langle \mathcal{F}|\Lambda|\mathcal{F}\rangle = 2 \int_{\mathbb{R}^3} d\mathbf{r} \, \frac{1}{Z_0} \mathbf{B}_{\text{re}}(\mathbf{r},t)^\dagger \mathbf{A}_{\text{re}}(\mathbf{r},t) - Z_0 \mathbf{D}_{\text{re}}(\mathbf{r},t)^\dagger \mathbf{C}_{\text{re}}(\mathbf{r},t). \quad (A13)$$

Equivalence with Equation (A6) is shown after considering that the definition of the typical real fields $\mathcal{X}(\mathbf{r},t)$ in Equation (6) implies that $\mathcal{X}(\mathbf{r},t) = 2\mathbf{X}_{\text{re}}(\mathbf{r},t)$. Equation (A6) is finally reached by substituting all the $\mathbf{X}_{\text{re}}(\mathbf{r},t)$ fields with $\mathcal{X}(\mathbf{r},t)/2$ in Equation (A13).

Appendix C. Expressions for Computing Circular Dichroism

Let us assume that a chiral molecule is located at point $\mathbf{r}$, and embedded in a possibly frequency-dispersive, homogeneous, isotropic, achiral, and non-magnetic medium with permittivity ϵ_m^ω, permeability $\mu_m^\omega = \mu_0$, impedance $Z^\omega = \sqrt{\mu_0/\epsilon_m^\omega}$, speed of light $c_\omega = 1/\sqrt{\mu_0 \epsilon_m^\omega}$. The wavenumber in such medium is $k_\omega = \omega/c_\omega$. In this Appendix, these frequency-dependent quantities are assumed to substitute their constant vacuum counterparts in all the equations in the main text.

The most common time-harmonic light-molecule interaction model is the 6 × 6 dipolarizability tensor, which relates the external electric and magnetic fields at the location of the molecule with the electric $[\mathbf{p}^\omega(\mathbf{r})]$ and magnetic $[\mathbf{m}^\omega(\mathbf{r})]$ dipoles induced by the fields in the molecule:

$$\begin{bmatrix} \mathbf{p}^\omega(\mathbf{r}) \\ \mathbf{m}^\omega(\mathbf{r}) \end{bmatrix} = \begin{bmatrix} \underline{\alpha}^\omega_{ee} & \underline{\alpha}^\omega_{em} \\ \underline{\alpha}^\omega_{me} & \underline{\alpha}^\omega_{mm} \end{bmatrix} \begin{bmatrix} \mathbf{E}^\omega(\mathbf{r}) \\ \mathbf{H}^\omega(\mathbf{r}) \end{bmatrix}. \quad (A14)$$

One of the most relevant techniques for chiral molecule sensing is Circular Dichroism (CD), which measures their differential absorption upon subsequent illumination with the two helicities. Assuming that the field scattered by the molecule, i.e. the field radiated by the induced dipoles in Equation (A14), is negligible with respect to the incident field, it is possible to write the rotationally averaged molecular differential absorption as [36].

$$CD(\mathbf{r}) = \int_{>0}^{\infty} d\omega\, \operatorname{Re}\{\alpha^\omega_{me}\} \frac{\omega c_\omega}{2} \left[|\mathbf{F}^\omega_+(\mathbf{r})|^2 - |\mathbf{F}^\omega_-(\mathbf{r})|^2 \right] = \int_{>0}^{\infty} d\omega\, \operatorname{Re}\{\alpha^\omega_{me}\} 2 c^2_\omega C^\omega(\mathbf{r}), \quad (A15)$$

where $\operatorname{Re}\{\alpha^\omega_{me}\}$ is the real part of the rotational average of $\underline{\alpha}^\omega_{me}$, $C^\omega(\mathbf{r})$ is the optical chirality density introduced by Tang and Cohen [37], and the second equality follows from ([16] Equation (5)) and Equation (4).

Besides the dipolarizability model in Equation (A14), other light-molecule interaction descriptions are possible. For example, the T-matrix of the molecule may be used. The T-matrix is a common object in physics and engineering, which is intrinsically able to include all the multipolar orders of the light-matter interaction, and allows to efficiently compute the coupled electromagnetic response of different objects in a systematic and rigorous way [38]. The conversion between the dipolarizability tensor and the T-matrix of the molecule up to the dipolar order is ([39] Equation (A15)):

$$\begin{bmatrix} \underline{\alpha}^\omega_{ee} & \underline{\alpha}^\omega_{em} \\ \underline{\alpha}^\omega_{me} & \underline{\alpha}^\omega_{mm} \end{bmatrix} = \frac{-i6\pi}{c_\omega Z_\omega k^3_\omega} \begin{bmatrix} \underline{T}^\omega_{NN} & iZ\underline{T}^\omega_{NM} \\ -ic_\omega \underline{T}^\omega_{MN} & c_\omega Z \underline{T}^\omega_{MM} \end{bmatrix}, \quad (A16)$$

where $N(M)$ refers to the electric(magnetic) character. We may now use this conversion to substitute $\operatorname{Re}\{\alpha^\omega_{me}\}$ in Equation (A15):

$$\begin{aligned} CD(\mathbf{r}) &= \int_{>0}^{\infty} d\omega\, \operatorname{Re}\{T^\omega_{MN}\}(3\pi c^3_\omega) \frac{|\mathbf{F}^\omega_+(\mathbf{r})|^2 - |\mathbf{F}^\omega_-(\mathbf{r})|^2}{\omega^2} \\ &= \int_{>0}^{\infty} d\omega\, \operatorname{Re}\{T^\omega_{MN}\}(3\pi c^3_\omega) \left[|\mathbf{V}^\omega_+(\mathbf{r})|^2 - |\mathbf{V}^\omega_-(\mathbf{r})|^2\right]. \end{aligned} \quad (A17)$$

In experimental measurements, a solution of chiral molecules is confined in its recipient, which defines a volume $\mathbb{D}$. Assuming uniform concentration of molecules over $\mathbb{D}$, the total CD signal can be computed by the volume integral of any of the expressions in Equation (A15) or Equation (A17) over $\mathbb{D}$. For example:

$$\begin{aligned} CD &= \rho \int_\mathbb{D} d\mathbf{r} \int_{>0}^{\infty} d\omega\, \operatorname{Re}\{T^\omega_{MN}\}(3\pi c^3_\omega) \frac{|\mathbf{F}^\omega_+(\mathbf{r})|^2 - |\mathbf{F}^\omega_-(\mathbf{r})|^2}{\omega^2} \\ &= \rho \int_{>0}^{\infty} d\omega\, \frac{\operatorname{Re}\{T^\omega_{MN}\}(3\pi c^3_\omega)}{\omega} \int_\mathbb{D} d\mathbf{r}\, \frac{|\mathbf{F}^\omega_+(\mathbf{r})|^2 - |\mathbf{F}^\omega_-(\mathbf{r})|^2}{\omega}, \end{aligned} \quad (A18)$$

where ρ is a constant that depends on the molecular concentration. In the last expression, we recognize one of the integrands from the average helicity in Equation (20).

References

1. Calkin, M.G. An Invariance Property of the Free Electromagnetic Field. *Am. J. Phys.* **1965**, *33*, 958. [CrossRef]
2. Zwanziger, D. Quantum Field Theory of Particles with Both Electric and Magnetic Charges. *Phys. Rev.* **1968**, *176*, 1489–1495. [CrossRef]
3. Deser, S.; Teitelboim, C. Duality transformations of Abelian and non-Abelian gauge fields. *Phys. Rev. D* **1976**, *13*, 1592–1597. [CrossRef]
4. Bialynicki-Birula, I.; Newman, E.T.; Porter, J.; Winicour, J.; Lukacs, B.; Perjes, Z.; Sebestyen, A. A note on helicity. *J. Math. Phys.* **1981**, *22*, 2530–2532. [CrossRef]
5. Bialynicki-Birula, I. Photon Wave Function. *Prog. Opt.* **1996**, *36*, 245–294.
6. Afanasiev, G.; Stepanovsky, Y. The helicity of the free electromagnetic field and its physical meaning. *Il Nuovo Cimento A (1971–1996)* **1996**, *109*, 271–279. [CrossRef]
7. Trueba, J.L.; Rañada, A.F. The electromagnetic helicity. *Eur. J. Phys.* **1996**, *17*, 141–144. [CrossRef]
8. Drummond, P.D. Dual symmetric Lagrangians and conservation laws. *Phys. Rev. A* **1999**, *60*, R3331–R3334. [CrossRef]
9. Coles, M.M.; Andrews, D.L. Chirality and angular momentum in optical radiation. *Phys. Rev. A* **2012**, *85*, 063810. [CrossRef]
10. Fernandez-Corbaton, I.; Zambrana-Puyalto, X.; Tischler, N.; Vidal, X.; Juan, M.L.; Molina-Terriza, G. Electromagnetic Duality Symmetry and Helicity Conservation for the Macroscopic Maxwell's Equations. *Phys. Rev. Lett.* **2013**, *111*, 060401. [CrossRef]
11. Cameron, R.P.; Barnett, S.M.; Yao, A.M. Optical helicity, optical spin and related quantities in electromagnetic theory. *New J. Phys.* **2012**, *14*, 053050. [CrossRef]
12. Bliokh, K.Y.; Bekshaev, A.Y.; Nori, F. Dual electromagnetism: helicity, spin, momentum and angular momentum. *New J. Phys.* **2013**, *15*, 033026. [CrossRef]
13. Cameron, R.P. On the 'second potential' in electrodynamics. *J. Opt.* **2013**, *16*, 015708. [CrossRef]
14. Nieto-Vesperinas, M. Optical theorem for the conservation of electromagnetic helicity: Significance for molecular energy transfer and enantiomeric discrimination by circular dichroism. *Phys. Rev. A* **2015**, *92*, 023813. [CrossRef]
15. Gutsche, P.; Poulikakos, L.V.; Hammerschmidt, M.; Burger, S.; Schmidt, F. Time-harmonic optical chirality in inhomogeneous space. In *Photonic and Phononic Properties of Engineered Nanostructures VI*; International Society for Optics and Photonics: San Francisco, CA, USA, 2016; Volume 9756, p. 97560X.
16. Fernandez-Corbaton, I.; Fruhnert, M.; Rockstuhl, C. Objects of Maximum Electromagnetic Chirality. *Phys. Rev. X* **2016**, *6*, 031013. [CrossRef]
17. Elbistan, M.; Horváthy, P.; Zhang, P.M. Duality and helicity: The photon wave function approach. *Phys. Lett. A* **2017**, *381*, 2375–2379. [CrossRef]
18. Andrews, D.L. Symmetries, Conserved Properties, Tensor Representations, and Irreducible Forms in Molecular Quantum Electrodynamics. *Symmetry* **2018**, *10*, 298. [CrossRef]
19. Vázquez-Lozano, J.E.; Martínez, A. Optical Chirality in Dispersive and Lossy Media. *Phys. Rev. Lett.* **2018**, *121*, 043901. [CrossRef]
20. Hanifeh, M.; Albooyeh, M.; Capolino, F. Optimally Chiral Electromagnetic Fields: Helicity Density and Interaction of Structured Light with Nanoscale Matter. *arXiv* **2018**, arXiv:1809.04117.
21. Crimin, F.; Mackinnon, N.; Götte, J.B.; Barnett, S.M. On the conservation of helicity in a chiral medium. *J. Opt.* **2019**, *21*, 094003. [CrossRef]
22. Guasti, M.F. Chirality, helicity and the rotational content of electromagnetic fields. *Phys. Lett. A* **2019**, *383*, 3180–3186. [CrossRef]
23. Poulikakos, L.V.; Dionne, J.A.; García-Etxarri, A. Optical Helicity and Optical Chirality in Free Space and in the Presence of Matter. *Symmetry* **2019**, *11*, 1113. [CrossRef]
24. Bernabeu, J.; Navarro-Salas, J. A Non-Local Action for Electrodynamics: Duality Symmetry and the Aharonov-Bohm Effect, Revisited. *Symmetry* **2019**, *11*, 1191. [CrossRef]
25. Nienhuis, G. Conservation laws and symmetry transformations of the electromagnetic field with sources. *Phys. Rev. A* **2016**, *93*, 023840. [CrossRef]
26. Gross, L. Norm Invariance of Mass-Zero Equations under the Conformal Group. *J. Math. Phys.* **1964**, *5*, 687–695. [CrossRef]

27. Mack, G.; Todorov, I. Irreducibility of the Ladder Representations of U(2, 2) when Restricted to the Poincaré Subgroup. *J. Math. Phys.* **1969**, *10*, 2078–2085. [CrossRef]
28. Mack, G. All unitary ray representations of the conformal group SU(2, 2) with positive energy. *Commun. Math. Phys.* **1977**, *55*, 1–28. [CrossRef]
29. Rose, M.E. *Elementary Theory of Angular Momentum*; Wiley: New York, NY, USA, 1957.
30. Jackson, J.D. *Classical Electrodynamics*; Wiley: New York, NY, USA, 1998.
31. Birula, I.B.; Birula, Z.B. *Quantum Electrodynamics by Iwo Białynicki-Birula and Zofia Białynicki-Birula*; Pergamon: Oxford, UK, 1975.
32. Cohen-Tannoudji, C.; Dupont-Roc, J.; Grynberg, G. *Photons and Atoms: Introduction to Quantum Electrodynamics*; Wiley: New York, NY, USA, 1989.
33. Messiah, A. *Quantum Mechanics*; Norh-Holland Publishing Company: Amsterdam, The Netherlands, 1961.
34. Bialynicki-Birula, I.; Bialynicka-Birula, Z. Beams of electromagnetic radiation carrying angular momentum: The Riemann-Silberstein vector and the classical-quantum correspondence. *Opt. Commun.* **2006**, *264*, 342–351. [CrossRef]
35. Duoandikoetxea, J.; Zuazo, J.D. *Fourier Analysis*; American Mathematical Soc.: Providence, RI, USA, 2001; Volume 29.
36. Graf, F.; Feis, J.; Garcia-Santiago, X.; Wegener, M.; Rockstuhl, C.; Fernandez-Corbaton, I. Achiral, Helicity Preserving, and Resonant Structures for Enhanced Sensing of Chiral Molecules. *ACS Photonics* **2019**, *6*, 482–491. [CrossRef]
37. Tang, Y.; Cohen, A.E. Optical Chirality and Its Interaction with Matter. *Phys. Rev. Lett.* **2010**, *104*, 163901. [CrossRef]
38. Mishchenko, M.I.; Zakharova, N.T.; Khlebtsov, N.G.; Videen, G.; Wriedt, T. Comprehensive thematic T-matrix reference database: A 2015–2017 update. *J. Quant. Spectrosc. Radiat. Transf.* **2017**, *202*, 240–246. [CrossRef]
39. Fernandez-Corbaton, I.; Rockstuhl, C.; Klopper, W. Computation of electromagnetic properties of molecular ensembles. *arXiv* **2018**, arXiv:1804.08085.

© 2019 by the authors. Licensee MDPI, Basel, Switzerland. This article is an open access article distributed under the terms and conditions of the Creative Commons Attribution (CC BY) license (http://creativecommons.org/licenses/by/4.0/).

Article

Role of Geometric Shape in Chiral Optics

Philipp Gutsche [1,2,*], **Xavier Garcia-Santiago** [3], **Philipp-Immanuel Schneider** [4], **Kevin M. McPeak** [5], **Manuel Nieto-Vesperinas** [6] **and Sven Burger** [2,4]

1. Freie Universität Berlin, Mathematics Institute, 14195 Berlin, Germany
2. Zuse Institute Berlin, Computational Nano Optics, 14195 Berlin, Germany; burger@zib.de
3. Karlsruhe Institute of Technology, Institut für Theoretische Festkörperphysik, 76131 Karlsruhe, Germany; xavier.garcia-santiago@kit.edu
4. JCMwave GmbH, 14050 Berlin, Germany; philipp.schneider@jcmwave.com
5. Department of Chemical Engineering, Lousiana State University, Baton Rouge, LA 70803, USA; kmcpeak@lsu.edu
6. Instituto de Ciencia de Materiales de Madrid, Consejo Superior de Investigaciones Científicas, 28049 Madrid, Spain; mnieto@icmm.csic.es
* Correspondence: gutsche@zib.de

Received: 11 November 2019; Accepted: 8 January 2020; Published: 13 January 2020

Abstract: The distinction of chiral and mirror symmetric objects is straightforward from a geometrical point of view. Since the biological as well as the optical activity of molecules strongly depend on their handedness, chirality has recently attracted high interest in the field of nano-optics. Various aspects of associated phenomena including the influences of internal and external degrees of freedom on the optical response have been discussed. Here, we propose a constructive method to evaluate the possibility of observing any chiral response from an optical scatterer. Based on solely the T-matrix of one enantiomer, planes of minimal chiral response are located and compared to geometric mirror planes. This provides insights into the relation of geometric and optical properties and enables identifying the potential of chiral scatterers for nano-optical experiments.

Keywords: optical chirality; mirror symmetry; helicity; optical scatterer

1. Introduction

It is usually a simple task to tell by eye whether an object is chiral or not: Achiral objects are superimposable onto their mirror image and, accordingly, they possess a mirror plane [1]. Recently, chiral scatterers have gained significant interest in nano-optics due to their potential to enhance the weak optical signal of chiral molecules [2–4]. Especially, the quantities of optical chirality and optical helicity as well as their relation to duality symmetry are subjects of current research [5]. The most established experimental technique in this field is the analysis of the circular dichroism (CD) spectrum which equals the differential energy extinction due to the illumination by right- and left-handed circularly polarized light [6].

In order to observe such chiral electromagnetic response, it seems to be obvious that geometrically chiral scatterers are required. However, it has been shown that extrinsic chirality, that is, a chiral configuration of the illumination and geometric parameters, yields comparable effects as intrinsically chiral objects [7]. By tuning the far-field polarization of the illumination, large chiral near-fields may even be generated in the viscinity of achiral objects [8]. In CD measurements, randomly oriented molecules are investigated, which can be classified by their T-matrix [9]. The latter has been used for quantifying the electromagnetic (e.m.) chirality, based on a novel definition of it [10].

However, the quantification of the geometric chirality is an elusive task [11] and even the unambiguous association of the terms right- and left-handed enantiomer of a general object is impossible [12]. Different coefficients attempting to rate the chirality of an object are based on the maximal overlap of two mirror images [13] as well as the Hausdorff distance [14]. The choice of a specific coefficient determines the most chiral object [15], that is, there is no natural choice for quantifying geometric chirality. This also holds for the various figures of merit estimating the e.m. chirality. Similar correlations between geometric and optical properties are investigated with respect to the non-sphericity of arbitrary scatterers [16].

In this study, we start by transferring the simple procedure of finding a mirror plane to optics. Such symmetries are present in different mathematical descriptions as for example, block structures in the Mueller scattering matrix [17]. Here, we analyse the T-matrix and its associated geometric mirror symmetries by employing translation and rotation theorems of vector spherical harmonics. We illustrate this concept with numerical simulations of an experimentally realized gold helix. Different quantifications of the e.m. chirality are compared. Furthermore, the symmetry planes found in the optical response by our method are correlated to those of geometric origin. It is shown that the complex optical response, including higher order multipoles, yields mirror planes in the T-matrix which are not directly related to geometric symmetries.

In the following, we would like to briefly introduce the theory behind the methods described in this study. Further information may be found in Supplementary Materials.

The most general description of an isolated optical scatterer is the well-known T-matrix [18]. It relates an arbitrary incident field with the scattered field caused by the scattering object. The optical response to *any* incident field is included in the T-matrix. Accordingly, the following analysis of T is independent of *specific* illumination parameters such as the direction, polarization and beam shape. The goal of this study is to obtain insights into illumination-*independent* symmetries of the scatterer.

Usually, both the incident as well as the scattered field are given in the basis of vector spherical harmonics for computations with the T-matrix [19] (see Supplementary Materials). Physically observable quantities such as the scattered energy, the absorption, as well as the flux of optical chirality are readily computed from T [9]. In numerical simulations, T may be computed with high accuracy [20]. Knowing the response of the left-handed object T_l enables the analytic computation of the response of its mirror image T_r:

$$T_r = \mathcal{M}_{xy}^{-1} T_l \mathcal{M}_{xy}, \tag{1}$$

where we choose mirroring in the xy-plane $\mathcal{M}_{xy}$ without loss of generality (see Supplementary Materials for further details on notation). Note that the terminology of right- T_r and left-handed T_l is ambiguous, as pointed out before, and may be interchanged.

Since we aim to investigate arbitrary mirror planes, we note that an arbitrary plane is given by the three spherical coordinates of its normal: the inclination Θ and the azimuthal angle Φ, as well as the distance d from the origin. We define the according transformation $R(\Theta, \Phi, d)$ acting on the object as

$$R(\Theta, \Phi, d) = \mathcal{T}(\Theta, \Phi, d) \mathcal{R}_z(\Phi) \mathcal{R}_y(\Theta), \tag{2}$$

where $\mathcal{T}(\Theta, \Phi, d)$ is the translation of the T-matrix in the direction given by the angles and the distance and $\mathcal{R}_z(\Phi)$ and $\mathcal{R}_y(\Theta)$ are the rotations around the z- and y-axis, respectively [21] (see Supplementary Materials).

For a geometrically achiral object (see Figure 1a) there exists at least one transformation $R(\Theta, \Phi, d)$ such that $T_l = R(\Theta, \Phi, d) T_r R^{-1}(\Theta, \Phi, d)$. On the other hand, the lack of a geometric mirror plane of a chiral object Figure 1b implies that there exists no such transformation and that T_l and T_r do not coincide for any set of transformation parameters (Θ, Φ, d). Note that this does not generally hold in the long wavelength limit, that is, the incident wavelength being much larger than the dimension of the scatterer, due to chiral dispersion [22].

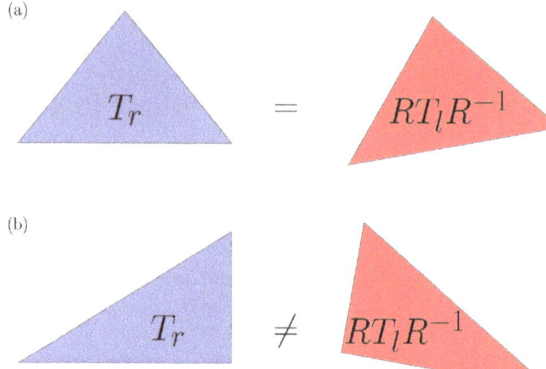

Figure 1. (**a**) The mirror image of an achiral object overlaps with its original after proper translations and rotations. This implies that the original T-matrix T_r coincides with T_l of the mirror object after the corresponding transformations R. (**b**) A chiral object and its mirror image are not congruent. If the object is much smaller than the incident wavelength, it usually exists a transformation R after which T_l and RT_rR^{-1} are equal. Note that the achiral isosceles triangle in (**a**) possess a mirror plane in 2D and that the asymmetric triangle in (**b**) is chiral only in 2D.

2. Results

For investigating the role of the geometric shape in nano-optics, it is of interest to identify those planes of highest symmetry of a chiral object: Although there is no mirror plane in a chiral object, a transformation may be identified in which the right- and left-handed T-matrices are closest to one another. Rating the closeness is done here by calculating the 2-norm of the difference of these two matrices. Accordingly, we introduce the coefficient χ_{TT} which minimizes the difference between the T-matrices of mirror images as

$$\chi_{TT} = \min_{(\Theta, \Phi, d)} \left\| T_l - R^{-1}(\Theta, \Phi, d) T_r R(\Theta, \Phi, d) \right\|_2. \tag{3}$$

This means that for the mirror plane corresponding to minimal parameters $(\Theta_{\min}, \Phi_{\min}, d_{\min})$ of (3), the optical responses of the two mirror images are as similar as possible. In other words, the mirror images are hardly distinguishable. For an achiral object χ_{TT} vanishes since there exists a transformation for which the mirror images are identical.

Obviously, the choice of the norm is not unique and other quantifications of similarity of the mirror images could be defined (see Supplementary Materials for the physical relevance of the 2-norm). A recently introduced coefficient χ_{SV} is, for example, based on the singular-value decomposition of the T-matrix in the helicity basis [10]. Alternatively, the angular-averaged differential energy extinction χ_{CD} due to illuminating with either right- or left-handed circularly polarized plane waves is experimentally accessible as the CD spectrum.

In order to exemplary introduce our formalism and compare it to previous work, we investigate a nano-optical device numerically. The finite element method is employed to accurately simulate the electromagnetic properties due to incident monochromatic light. Within this study we use the commercial FEM package JCMsuite [23]. In postprocessing, the T-matrix is computed by decomposing the scattered field into vector spherical wave functions [20] from illumination with 150 plane waves with randomly chosen parameters (see Supplementary Materials).

In Figure 2, we compare simulations of the aforementioned three coefficients quantifying the e.m. chirality for a gold helix as realized experimentally [24]. The helix is constructed on the surface of a cylinder with height 230nm and radius 60nm (see Supplementary Materials). The CD spectrum χ_{CD} shows zero values at incident wavelengths of $\lambda = 615$ nm and $\lambda = 1070$ nm. If only these wavelengths were analyzed, one could draw the conclusion that an achiral object is investigated. This contradicts the goal of this study to obtain insights into illumination-*independent* symmetries of the scatterer—for illuminations with $\lambda = 615$ nm and $\lambda = 1070$ nm, the scatterer seems to be geometrically achiral which is obviously not the case. Nevertheless, CD makes the chiral geometric nature of the helix visible as a maximum at 823 nm and a minimum at 1452 nm of smaller amplitude. For a helix with an opposite twist—that is, the mirror image—the roles of the extrema are interchanged.

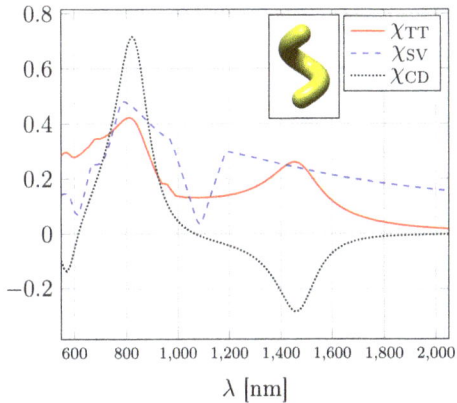

Figure 2. Chiral response of a gold nano-helix depending on the incident wavelength λ. The angular averaged differential extinction of circularly polarized plane waves χ_{CD} (black dotted line) vanishes at 615 nm and 1070 nm which could be interpreted as achirality of the studied object. The electromagnetic chirality coefficient χ_{SV} (dashed blue line) is based on the singular values of the *T*-matrix in the helicity basis. Values below 0.1 at 610 nm and 1085 nm indicate nearly achiral optical response. However, the minimal difference χ_{TT} (red solid line) between T_r and RT_lR^{-1} reveals that the helix is chiral at all wavelengths. Its maxima correspond to those of χ_{CD} and are, hence, observable.

On the other hand, the coefficient χ_{SV} is normalized by the average interaction strength of the *T*-matrix at each wavelength. This yields a fairly flat spectrum with two narrow minima below 0.1 at the two λ for which $\chi_{CD} = 0$. These minima are not present in the minimized χ_{TT} introduced in (3). However, the maxima of this latter coefficient are in accordance with the experimentally observable CD extrema (χ_{CD}). In the long wavelength regime, all three coefficients tend to zero as expected for point-like particles due to vanishing off-diagonal elements in the *T*-matrix.

The minimization in the three-dimensional parameter space in (3) is carried out using Bayesian optimization [25] (see Supplementary Materials). Since the shape of the minimized function highly depends on the actual object, the Bayesian approach is well suited for finding a global minimum. The parameters $(\Theta_{min}, \Phi_{min}, d_{min})$ of the optimized value are related to geometric mirror planes. In Figure 3a, the planes following from the respective transformation $R(\Theta_{min}, \Phi_{min}, d_{min})$ of the *xy*-plane are plotted for all incident wavelengths from 550 nm to 2.05 µm. The inclination Θ and azimuthal angle Φ are given in the shown coordinate system which is centered at the centroid of the helix.

We identify three distinct classes shown in blue, red and green. These correspond to planes which are parallel and perpendicular to the helix axis, as well as tilted by a small angle Θ from the horizontal position, respectively. The dark grey plane corresponds to the minimal geometric parameters which will be explained in the following paragraphs. Details on the optimization such as challenging flat behaviour for translations from the centroid and, on the obtained minimizing parameters, are given in Supplementary Materials. Note that the minimization required to obtain the illumination-independent coefficient χ_{TT} involves significantly higher numerical effort than the simple averaging for χ_{CD} for which most information contained in *T* is ignored.

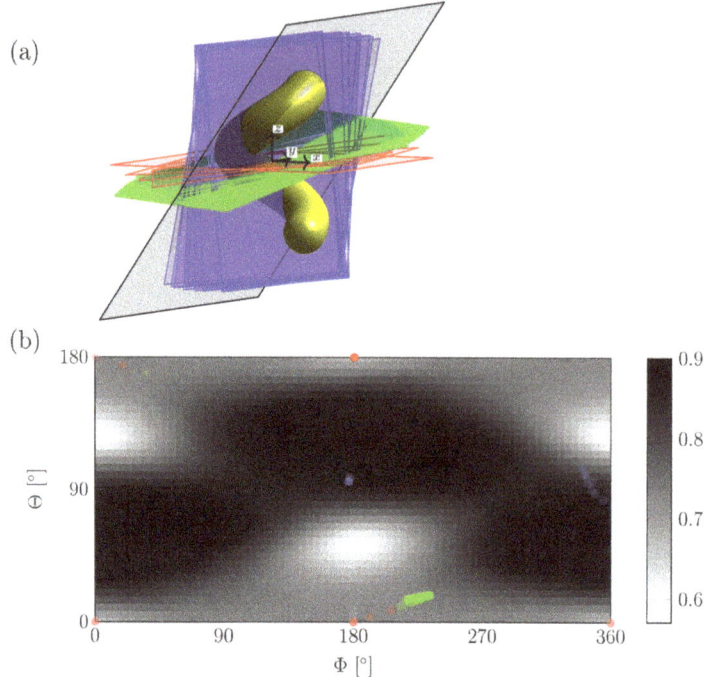

Figure 3. (a) Transformed xy-planes (blue, red, green) corresponding to minimal χ_{TT} computed from T-matrix of the gold helix (yellow). Planes for all incident wavelenghts $\lambda \in [550, 2050]$ nm are shown. The dark grey plane corresponds to minimal χ_{GE}. (b) Geometric chiral coefficient $\chi_{GE}(\Theta, \Phi)$ for the helix and its mirror image which is rotated around the centroid (grey colormap). The minimal value of 0.57 belongs to the dark grey plane in Figure 3a. Angles of the colored planes are shown by circles.

Next, we compare the findings on the symmetry based on the optical T-matrix to those stemming from purely geometric properties. As discussed previously, there is no coefficient which unambiguously rates the geometric chirality of an object. We choose a coefficient χ_{GE} based on the overlap of the left- O_l and right-handed $O_r(\Theta, \Phi, d)$ object, where the latter results from mirroring O_l at the xy-plane and transformation with (Θ, Φ, d). Namely, the volume V of the overlap is compared to the volume of the object [13] (see Supplementary Materials):

$$\chi_{GE}(\Theta, \Phi, d) = 1 - \frac{V(O_l \cap O_r(\Theta, \Phi, d))}{V(O_l)}. \tag{4}$$

This coefficient vanishes for achiral objects as required for a degree of chirality [14].

Figure 3b displays the geometric chirality coefficient $\chi_{GE}(\Theta, \Phi, 0)$ for planes rotated around the centroid of the helix as a grey colourmap. Dark regions with large values of χ_{GE} indicate a vanishing overlap between the two mirrored helices. Note that for large distances to the origin $d \to \infty$, the mirror images do not overlap and $\chi_{GE} = 1$. However, this is always possible no matter if the object is chiral or not. As in the case of χ_{TT}, the parameter points of interest of $\chi_{GE}(\Theta, \Phi, d)$ are those corresponding to a minimum: The minimum 0.57 in Figure 3b occurs at $(180°, 55°)$ and $(0°, 125°)$ which show the intrinsic chiral property of the investigated helix. These two minima are equivalent since a finite helix is C_2 symmetric. The corresponding transformed xy-plane is shown in dark grey in Figure 3a.

Alongside the geometric coefficient χ_{GE}, the planes identified for the minimized T-matrix difference are shown as colored circles in Figure 3b. The colors (red, blue and green) of these circles are the same colors used for the planes, that is, a direct comparison of the angle parameters is possible. As seen, the planes are ranked according to their Θ values: The perpendicular class 1 (blue) has $\Theta \in [83, 105]°$ The flat planes belonging to class 2 (red) show $\Theta \in [0, 8.5]°$ and $\Theta \in [174, 180]°$ and the tilted class 3 (green) has $\Theta \in [10, 19]°$ and $\Theta = 170°$.

3. Discussion

None of the three optical symmetry planes is directly related to the geometric mirror plane of the helix. However, Figure 3b enables the comparison of geometric and optical symmetries. In order to further analyze the optical response, we show the wavelength-dependent classification of the symmetry planes on top of Figure 4. The three classes correspond to sharply separated wavelength ranges: Class 1 is valid for $\lambda \in [550, 680]$ nm. For larger wavelengths $\lambda \in [680, 1025]$ nm, the T-matrix possesses the symmetry according to planes of class 2. Finally, in the long wavelength regime ($\lambda \in [1025, 2050]$ nm), the symmetry is in class 3.

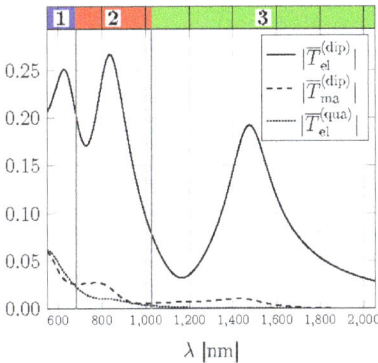

Figure 4. Wavelength-dependent classification of symmetry planes of T-matrix (top). Absolute value of averaged diagonal T-matrix entries corresponding to induced electric dipoles (solid), magnetic dipoles (dashed) and electric quadrupoles (dotted). The classes 3 (green), 2 (red) and 1 (blue) belong to decreasing wavelengths. Changes in symmetry of T are due to higher order multipoles.

The analysis in Figure 3b suggests that class 3 (green) is the closest one to the geometric mirror plane. This is further strengthened by the full angular spectrum of the optical chirality coefficient χ_{TT} (see Supplementary Materials). Accordingly, we find that the optical response is dominated by the geometric shape for long wavelengths. Obviously, the optics is dominated by the electric dipole moment in this regime which is also shown in Figure 4. Here, the mean of the diagonal entries of submatrices of the T-matrix are shown. These are proportional to the electric and magnetic dipole moments as well as to the electric quadrupole moments.

The three symmetry classes of the T-matrix occur close to three electric dipole peaks ($\lambda = 623$, 833, and 1473 nm) and are influenced by the anisotropy of the T-matrix. Truly chiral behaviour, as observed here, however, originates not from anisotropy but from coupling between electric and magnetic multipoles [26]. In Supplementary Materials, we elaborate on the complex interplay between these different contributions in the dipolar limit. Here, we limit the discussion to the main aspects of different multipolar contributions.

For large wavelengths with symmetry of class 3, the electric dipoles are much larger then any other induced multipole. In the intermediated regime of symmetry class 2, the magnetic dipole moment significantly increases. For short wavelengths with planes of class 1, the electric quadrupole moment is stronger than the magnetic dipole moment which yields the change in the optical symmetry. Higher order multipoles including mixed electric-magnetic moments are depicted in Supplementary Materials, in which it is shown that the dominant dipolar moments contribute additionally to the variation of mirror planes. This elaborated study of multipolar resonances underlines again that the chiral response deviates from expectations due to a purely geometrical analysis of the scatterer.

4. Conclusions

In summary, we have introduced a method to obtain geometric mirror planes from the optical T-matrix of a scattering object. Accordingly, the optical effects of geometric structures such as metamaterials are analyzed [27]. We applied the procedure to an isolated gold helix and found correlations between the symmetry of its geometric shape and those of the optical response in the long wavelength regime. On the one hand, this confirms the expectation that instrinsic geometric chirality is directly related to an optically chiral response. On the other hand, for shorter wavelengths where higher multipoles are induced, mirror planes derived from the T-matrix do not coincide with the geometric mirror plane. This implies light-matter interactions whose symmetry cannot be explained simply by geometric chirality. Our method can be applied to all isolated scattering objects being chiral as the helix or achiral (see Supplementary Materials). It constructively identifies geometric planes of mirror symmetry in their optical response. This approach provides the basis for a detailed analysis of correlations between structural and spectral properties of nano-optical scatterers.

Supplementary Materials: The following are available online at http://www.mdpi.com/2073-8994/12/1/158/s1 and contain detailed information on T-matrix formalism, electromagnetic and geometric chirality coefficients, multipolar analysis, geometric model and optimization as well as the analysis of an achiral scatterer with our method.

Author Contributions: Conceptualization, P.G. and K.M.M.; methodology, P.G. and P.-I.S.; software, X.G.-S. and P.-I.S.; resources M.N.-V. and K.M.M.; writing—original draft preparation, P.G.; writing—review and editing, X.G.-S., P.-I.S., K.M.M., M.N.-V. and S.B.; supervision, M.N.-V. and S.B. All authors have read and agreed to the published version of the manuscript.

Funding: We acknowledge support from Freie Universität Berlin through the Dahlem Research School. This research was funded by the European Union's Horizon 2020 research and innovation programme under the Marie Sklodowska-Curie grant number 675745. This research was funded by the EMPIR programme co-financed by the Participating States and from the European Union's Horizon 2020 research and innovation programme under grant number 17FUN01 (BeCOMe). M. Nieto-Vesperinas acknowledges Spanish Ministerio de Ciencia, Innovación y Universidades, grants FIS2014-55563-REDC, FIS2015-69295-C3-1-P, and PGC2018-095777-B-C21.

Acknowledgments: We thank Martin Hammerschmidt for in-depth discussions on several topics.

Conflicts of Interest: The authors declare no conflict of interest.

References

1. Kelvin, W.T.B. *Baltimore Lectures on Molecular Dynamics and the Wave Theory of Light*; CJ Clay and Sons: Baltimore, MD, USA, 1904.
2. Tang, Y.; Cohen, A.E. Optical chirality and its interaction with matter. *Phys. Rev. Lett.* **2010**, *104*, 163901. [CrossRef] [PubMed]
3. Nieto-Vesperinas, M. Optical theorem for the conservation of electromagnetic helicity: Significance for molecular energy transfer and enantiomeric discrimination by circular dichroism. *Phys. Rev. A* **2015**, *92*, 023813. [CrossRef]
4. McPeak, K.M.; van Engers, C.D.; Bianchi, S.; Rossinelli, A.; Poulikakos, L.V.; Bernard, L.; Herrmann, S.; Kim, D.K.; Burger, S.; Blome, M.; et al. Ultraviolet Plasmonic Chirality from Colloidal Aluminum Nanoparticles Exhibiting Charge-Selective Protein Detection. *Adv. Mater.* **2015**, *27*, 6244–6250. [CrossRef] [PubMed]
5. Poulikakos, L.V.; Dionne, J.A.; García-Etxarri, A. Optical Helicity and Optical Chirality in Free Space and in the Presence of Matter. *Symmetry* **2019**, *11*, 1113. [CrossRef]
6. Bohren, C.F.; Huffman, D.R. *Absorption and Scattering of Light by Small Particles*; John Wiley & Sons: Hoboken, NJ, USA, 1983.
7. Plum, E.; Fedotov, V.; Zheludev, N. Optical activity in extrinsically chiral metamaterial. *Appl. Phys. Lett.* **2008**, *93*, 191911. [CrossRef]

8. Kramer, C.; Schäferling, M.; Weiss, T.; Giessen, H.; Brixner, T. Analytic optimization of near-field optical chirality enhancement. *ACS Photonics* **2017**, *4*, 396–406. [CrossRef] [PubMed]
9. Gutsche, P.; Nieto-Vesperinas, M. Optical Chirality of Time-Harmonic Wavefields for Classification of Scatterer. *Sci. Rep.* **2018**, *8*, 9416. [CrossRef]
10. Fernandez-Corbaton, I.; Fruhnert, M.; Rockstuhl, C. Objects of maximum electromagnetic chirality. *Phys. Rev. X* **2016**, *6*, 031013. [CrossRef]
11. Fowler, P.W. Quantification of chirality: Attempting the impossible. *Symmetry Cult. Sci.* **2005**, *16*, 321–334.
12. Efrati, E.; Irvine, W.T. Orientation-dependent handedness and chiral design. *Phys. Rev. X* **2014**, *4*, 011003. [CrossRef]
13. Gilat, G. On quantifying chirality-obstacles and problems towards unification. *J. Math. Chem.* **1994**, *15*, 197–205. [CrossRef]
14. Buda, A.B.; Mislow, K. A Hausdorff chirality measure. *J. Am. Chem. Soc.* **1992**, *114*, 6006–6012. [CrossRef]
15. Rassat, A.; Fowler, P.W. Is there a "most chiral tetrahedron"? *Chemistry* **2004**, *10*, 6575–6580. [CrossRef] [PubMed]
16. Romanov, A.V.; Konokhova, A.I.; Yastrebova, E.S.; Gilev, K.V.; Strokotov, D.I.; Maltsev, V.P.; Yurkin, M.A. Sensitive detection and estimation of particle non-sphericity from the complex Fourier spectrum of its light-scattering profile. *J. Quant. Spectrosc. Radiat. Transf.* **2019**, *235*, 317–331. [CrossRef]
17. Yurkin, M.A. Symmetry relations for the Mueller scattering matrix integrated over the azimuthal angle. *J. Quant. Spectrosc. Radiat. Transf.* **2013**, *131*, 82–87. [CrossRef]
18. Mishchenko, M.; Travis, L.; Lacis, A. *Scattering, Absorption, and Emission of Light by Small Particles*; Cambridge University Press: Cambridge, UK, 2002.
19. Jackson, J.D. *Classical Electrodynamics*, 3rd ed.; John Wiley and Sons: Hoboken, NJ, USA, 1998.
20. Garcia Santiago, X.; Hammerschmidt, M.; Burger, S.; Rockstuhl, C.; Fernandez Corbaton, I.; Zschiedrich, L. Decomposition of scattered electromagnetic fields into vector spherical wave functions on surfaces with general shapes. *Phys. Rev. B* **2019**, *99*, 045406. [CrossRef]
21. Stein, S. Addition theorems for spherical wave functions. *Q. Appl. Math.* **1961**, *19*, 15–24. [CrossRef]
22. Lindell, I.V.; Sihvola, A.H. *Electromagnetic Wave in Chiral and Bi-Isotropic Media*; Artech House: London, UK, 1994.
23. Pomplun, J.; Burger, S.; Zschiedrich, L.; Schmidt, F. Adaptive finite element method for simulation of optical nano structures. *Phys. Status Solidi B* **2007**, *244*, 3419–3434. [CrossRef]
24. Wozniak, P.; De Leon, I.; Höflich, K.; Haverkamp, C.; Christiansen, S.; Leuchs, G.; Banzer, P. Chiroptical response of a single plasmonic nanohelix. *Opt. Express* **2018**, *26*, 19275–19293. [CrossRef]
25. Schneider, P.I.; Santiago, X.G.; Rockstuhl, C.; Burger, S. Global optimization of complex optical structures using Bayesian optimization based on Gaussian processes. *Proc. SPIE* **2017**, *10335*, 103350O.
26. Zambrana-Puyalto, X.; Bonod, N. Tailoring the chirality of light emission with spherical Si-based antennas. *Nanoscale* **2016**, *8*, 10441–10452. [CrossRef] [PubMed]
27. Nakata, Y.; Urade, Y.; Nakanishi, T. Geometric Structure behind Duality and Manifestation of Self-Duality from Electrical Circuits to Metamaterials. *Symmetry* **2019**, *11*, 1336. [CrossRef]

© 2020 by the authors. Licensee MDPI, Basel, Switzerland. This article is an open access article distributed under the terms and conditions of the Creative Commons Attribution (CC BY) license (http://creativecommons.org/licenses/by/4.0/).

Article

Geometric Structure behind Duality and Manifestation of Self-Duality from Electrical Circuits to Metamaterials

Yosuke Nakata [1,*,†], Yoshiro Urade [2] and Toshihiro Nakanishi [3]

1. Research Center for Advanced Science and Technology, The University of Tokyo, Meguro-ku, Tokyo 153-8904, Japan
2. Center for Emergent Matter Science, RIKEN, Wako, Saitama 351-0198, Japan; yoshiro.urade@riken.jp
3. Department of Electronic Science and Engineering, Kyoto University, Kyoto 615-8510, Japan; t-naka@kuee.kyoto-u.ac.jp
* Correspondence: nakata@ee.es.osaka-u.ac.jp
† Current address: Graduate School of Engineering Science, Osaka University, Toyonaka, Osaka 560-8531, Japan.

Received: 31 August 2019; Accepted: 14 October 2019; Published: 28 October 2019

Abstract: In electromagnetic systems, duality is manifested in various forms: circuit, Keller–Dykhne, electromagnetic, and Babinet dualities. These dualities have been developed individually in different research fields and frequency regimes, leading to a lack of unified perspective. In this paper, we establish a unified view of these dualities in electromagnetic systems. The underlying geometrical structures behind the dualities are elucidated by using concepts from algebraic topology and differential geometry. Moreover, we show that seemingly disparate phenomena, such as frequency-independent effective response, zero backscattering, and critical response, can be considered to be emergent phenomena of self-duality.

Keywords: duality; self-duality; Poincaré duality; circuit duality; Keller–Dykhne duality; electromagnetic duality; Babinet's principle; constant-resistance circuit; zero backscattering; critical response; self-complementary antenna; metamaterials; metasurfaces

1. Introduction

Duality is an indispensable concept in mathematics, physics, and engineering. It relates two seemingly different systems in a nontrivial manner and facilitates deeper insight into the underlying structures behind the relevant physical theories. A duality transformation converts an object to its dual counterpart. Performing duality transformations twice brings a system back to its original state. Advantageously, duality transformation can sometimes convert a difficult problem into a more tractable one. Moreover, if a system is dual to itself, namely self-dual, some special characteristics could be expected. To name a few, self-duality characterizes a critical probability and temperature for percolation on graphs [1,2] and two-dimensional Ising models [3], respectively.

For electromagnetic systems, duality appears in various forms, such as dual circuits [4,5], Keller–Dykhne duality [6–8], electromagnetic duality [9–11], input–impedance duality for antennas [12,13], and Babinet's principle [14–26]. These dualities have been individually developed in various research fields such as electrical engineering, radio-frequency engineering, and photonics because the relevant frequency spectra broadly range from direct-current to the optical regime. Despite the long history of dualities in electromagnetic systems, they have not been sufficiently discussed from a unified perspective. In particular, universal geometrical structures behind the dualities are still mathematically unclear in electromagnetic systems.

Recently, engineered artificial composites called metamaterials have been attracting much attention due to their exotic electromagnetic properties [27]. In particular, two-dimensional metamaterials are called metasurfaces and are intensively studied as ultrathin functional devices [28–30]. The operating frequency of metamaterials has been extended from the microwave to the optical band, and versatile design principles over extensively wide frequency spectra are in high demand. For example, circuit designs and nano-optics were integrated to establish a universal strategy to design metamaterials [31]. Over the past few decades, duality has been assuming an important role in the research and development of metamaterials. To name a few such cases, duality has been leveraged to design complementary metasurfaces [32–35], self-complementary metasurfaces [36–42], critical metasurfaces [43–52], maximally chiral metamaterials [53], and self-dual metamaterials with zero backscattering [54] or helicity conservation [55–57]. In such situations, the extensive target frequencies of metamaterials require a comprehensive understanding of duality in a unified manner.

In this paper, we establish a unified perspective which synthesizes dualities appearing in electromagnetic systems. The underlying geometrical structures hidden behind dualities are uncovered through Poincaré duality between circuit theory to optics. To this end, we introduce inner and outer orientations of geometrical objects. We stress the importance of these two different kinds of orientations because it is sometimes overlooked in primary electromagnetism. The correspondence relationship among various dualities in electromagnetic systems is thoroughly discussed. Moreover, we comprehensively show that self-duality manifests as frequency-independent effective response, zero backscattering, and criticality in electromagnetic systems. As a whole, we attempted to keep the paper pictorial as much as possible in order to easily grasp the important concepts for duality.

In Section 2, we start with discussions of electrical circuits. Although electrical circuits are simplified and idealized systems, they include the essence of duality. To clearly see duality in circuit theory, we provide an algebraic-topological framework for electrical circuits. Based on the established framework, we strictly formulate circuit duality through Poincaré duality. Furthermore, we explain that the effective response of a self-dual circuit is automatically derived from its self-duality. In Section 3, we discuss zero backscattering for waves in self-dual circuits in terms of impedance matching. Section 4 generalizes circuit duality to Keller–Dykhne duality for a continuous two-dimensional resistive sheet. We show that critical behavior can appear for a self-dual sheet. To extract the essence of the duality in continuous systems, differential forms are utilized. We see that Keller–Dyhkne duality exactly corresponds to circuit duality through discretization. In Section 5, we introduce electromagnetic duality to Maxwell's electromagnetic theory. Electromagnetic duality in a four-dimensional spacetime has a similar structure to Keller–Dykhne duality in a two-dimensional plane. Section 6 introduces duality for wave radiation and scattering by a two-dimensional sheet. Combining electromagnetic duality and mirror symmetry of the sheet, we establish Babinet duality. Babinet duality induces the critical response of a metallic checkerboard-like sheet. Finally, Section 7 summarizes the unified perspective for dualities appearing in electromagnetic systems and gives a conclusion.

2. Circuit Theory

In this section, we establish a comprehensive formulation of circuit theory based on algebraic topology. Within the proposed framework that makes use of Poincaré duality, we unveil the geometrical structure of circuit duality. Next, we discuss the resultant duality in effective response and show that a specific response is automatically guaranteed under self-duality. Compared to the previous algebraic topological formulation of the circuit theory [58–60], we clarify the importance of inner and outer orientations of geometrical objects for circuit duality. Our approach gives a succinct mathematical description of circuit duality, and an amalgamation of circuit theory and algebraic topology yields various benefits in this interdisciplinary area.

2.1. Duality between Current and Voltage

■ Circuits as Graphs

A circuit is considered as a network of interconnected components. The interconnection relation can be represented by a *graph* $\mathcal{G} = (\mathcal{N}, \mathcal{E})$ with a set $\mathcal{N} = \{n_i | i = 1, 2, \cdots, |\mathcal{N}|\}$ expressing the totality of nodes and a set $\mathcal{E} = \{e_i | i = 1, 2, \cdots, |\mathcal{E}|\}$ representing the totality of directed edges (an ordered 2-element subset of $\mathcal{N}$), where $|\mathcal{S}|$ is the number of elements of a set $\mathcal{S}$. Elements in $\mathcal{N}$ and $\mathcal{E}$ are called 0- and 1-cells, respectively. Circuit elements such as voltage sources or resistors are placed along the graph edges. In this section, we mainly focus on circuits with resistors, voltage, and current sources. Figure 1 shows an example of a circuit and the corresponding graph.

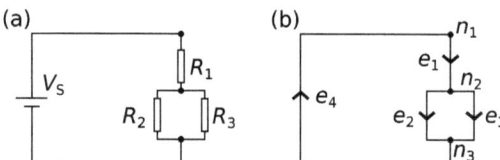

Figure 1. Example of (**a**) a circuit and (**b**) its corresponding graph, which represents the circuit interconnection.

■ Chains

Consider a current distribution with I^i flowing along each edge $e_i \in \mathcal{E}$. Here, e_i is directed and I^i flows in the direction of e_i. A negative I^i represents a current flow in the reverse direction of e_i. This current distribution can be represented by a formal sum $I = \sum_{i=1}^{|\mathcal{E}|} I^i e_i$, where $\mathcal{E}$ generates a vector space over real numbers $\mathbb{R}$. In a strict sense, we should consider $\mathbb{R}A := \{sA | s \in \mathbb{R}\}$ with the unit of ampere A instead of $\mathbb{R}$ for a current distribution, but we omitted the unit to avoid notation complexity. A vector spanned by elements in $\mathcal{E}$ is called a *1-chain*. The vector space composed of all 1-chains is denoted by C_1. Similarly, we can define a *0-chain* as $\sum_{i=1}^{|\mathcal{N}|} q^i n_i$ with $q^i \in \mathbb{R}$, and introduce a vector space C_0 over $\mathbb{R}$ as the totality of 0-chains.

■ Boundary Operator

Now, we extract the connection information regarding $\mathcal{G}$. Consider an edge e directing from a node n_1 to n_2. As shown in Figure 2, the boundary of e is given by $\partial e = n_2 - n_1$. Extending the definition linearly, we can introduce a linear *boundary operator* $\partial : C_1 \to C_0$. The matrix representation of ∂ is denoted by $\Delta^i{}_j$, called the *incidence matrix*. For the previous example in Figure 1b, we obtain the following matrix representation of ∂:

$$\partial [e_1 \ e_2 \ e_3 \ e_4] := [\partial e_1 \ \partial e_2 \ \partial e_3 \ \partial e_4] = [n_1 \ n_2 \ n_3] \Delta \tag{1}$$

with

$$\Delta = [\Delta^i{}_j] = \begin{bmatrix} -1 & 0 & 0 & 1 \\ 1 & -1 & -1 & 0 \\ 0 & 1 & 1 & -1 \end{bmatrix}. \tag{2}$$

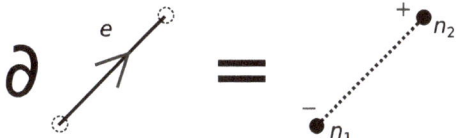

Figure 2. Action of the boundary operator. Here, we have $\partial e = n_2 - n_1$.

■ **Kirchhoff's Current Law**

To understand the physical meaning of ∂, we consider ∂I for a current distribution $I \in C_1$. The coefficient of ∂I for $n \in \mathcal{N}$ represents the net inflow of the current: the current inflow to n minus the current outflow from n. Therefore, the boundary operator can be used to express Kirchhoff's current law (KCL), which states that net current inflows at each node are zero. KCL restricts current distribution to a linear subspace $Z_1 = \ker \partial = \{c \in C_1 | \partial c = 0\}$. For Equation (2), we have $\operatorname{rank} \Delta = 2$, and $\dim Z_1 = \dim C_1 - \operatorname{rank} \Delta = 2$ from the rank–nullity theorem [61]. The basis of Z_1 is given by $\{m_1 = e_1 + e_2 + e_4,\ m_2 = e_2 - e_3\}$. Here, m_1 and m_2 are closed loops called *meshes*. Generally, we can construct a basis of Z_1 with meshes [60].

■ **Cochains**

Next, we represent a voltage distribution in geometric terms. For a finite-dimensional vector space U, we can define its dual space $U^* = \{\alpha : U \to \mathbb{R}\}$, where α is a linear map and U^* is a vector space with $\dim U^* = \dim U$. An element $\alpha \in U^*$ can be interpreted as an apparatus which measures a vector $u \in U$ and yields $\alpha(u)$. For a basis $\{u_1, u_2, \cdots, u_m\}$ in U, we can define a dual basis $\{u^1, u^2, \cdots, u^m\}$ in U^* satisfying $u^i(u_j) = \delta^i_j$, where the Kronecker delta δ^i_j is 1 if $i = j$; otherwise, 0. For a vector $v = \sum_{i=1}^m v^i u_i \in U$ with $v^i \in \mathbb{R}$, u^i extracts the component with respect to u_i as $u^i(v) = v^i$.

Along an edge, we can calculate power consumption as a real scalar equal to the voltage multiplied by the current. The total power consumption in the circuit is the sum of power consumption over all edges (power generation is represented by negative power consumption). Therefore, a voltage distribution is considered as $V \in C^1 = (C_1)^*$ yielding total power $V(I)$ for $I \in C_1$. An element in C^1 is called a *1-cochain*. In C^1, we have a dual basis $\{e^i | i = 1, 2, \cdots, |\mathcal{E}|\}$. Then, we can express $V(I) = \sum_{i=1}^{|\mathcal{E}|} V_i I^i$ for $V = \sum_{i=1}^{|\mathcal{E}|} V_i e^i$ and $I = \sum_{i=1}^{|\mathcal{E}|} I^i e_i$. We also write $V(I)$ as $\int_I V$ to stress the analogy to the theory of continuous fields. A 0-cochain $\varphi \in C^0 = (C_0)^*$ also acts for a 0-chain $b \in C_0$ as $\int_b \varphi = \varphi(b)$.

■ **Kirchhoff's Voltage Law**

Now, we reframe Kirchhoff's voltage law (KVL) in our geometric approach. KVL states that the sum of voltages along any loop must be zero. Then, a voltage distribution $V \in C^1$ must satisfy $\int_I V = 0$ for all $I \in Z_1$ (also known as Tellegen's theorem) because I is generated from mesh currents. To concisely express KVL, we define a null space $(U')^\perp$ for a linear subspace U' in U as $(U')^\perp = \{\alpha \in U^* | \alpha(u) = 0 \text{ for all } u \in U'\}$. The dimension of the null space is given by $\dim (U')^\perp = \dim U - \dim U'$. The relation between U' and $(U')^\perp$ is schematically depicted in Figure 3. By using the concept of a null space, KVL is clearly rewritten to restrict voltage distribution to $(Z_1)^\perp$.

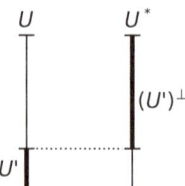

Figure 3. Relation between U' and $(U')^\perp$.

■ Dual Operators

Next, we want to define a dual boundary operator for ∂. Let U and W be finite-dimensional vector spaces, and consider a linear map $f : U \to W$. Then, a *dual map* $f^* : W^* \to U^*$ is defined as $[f^*(\alpha)](u) := \alpha[f(u)]$ for $\alpha \in W^*$ and all $u \in U$. Let $\{u_1, u_2, \cdots, u_l\}$ and $\{w_1, w_2, \cdots, w_m\}$ be the bases of U and W, respectively. We have $f^*(w^i) = \sum_{j=1}^{l}[f^*(w^i)](u_j)u^j = \sum_{j=1}^{l} w^i[f(u_j)]u^j = \sum_{j=1}^{l} u^j M^i{}_j$ with the matrix representation of f as $M^i{}_j = w^i[f(u_j)]$. This shows that the matrix representation of the dual map is the transpose of the matrix representation of the original map. Using the concept of the dual map, we can define a *coboundary* operator $d = \partial^*$ to satisfy $\int_{\partial c} \varphi = \int_c d\varphi$ for all $c \in C_1$ and $\varphi \in C^0 = (C_0)^*$. For a dual basis $\{n^i\} \subset C^0$ obtained from 0-cells $\mathcal{N} = \{n_i\}$, we have $dn^i = \sum_{j=1}^{|\mathcal{E}|} e^j \Delta^i{}_j$ with the incidence matrix $\Delta^i{}_j$. For the example of Figure 1b, we have

$$d[n^1 \ n^2 \ n^3] := [dn^1 \ dn^2 \ dn^3] = [e^1 \ e^2 \ e^3 \ e^4] \begin{bmatrix} -1 & 1 & 0 \\ 0 & -1 & 1 \\ 0 & -1 & 1 \\ 1 & 0 & -1 \end{bmatrix}. \qquad (3)$$

■ Potential and Kirchhoff's Voltage Law

Finally, we discuss a relation between KVL and a potential. KVL was formulated to state that the voltage drop along any loop is zero, but how is this statement related to the existence of a potential? To see this, we start from a general statement. Consider a linear map $f : U \to W$. The image and kernel of linear maps f and f^* are related through $(\ker f)^\perp = \operatorname{im} f^*$ and $(\operatorname{im} f)^\perp = \ker f^*$ [61]. The proof of the first statement is as follows. First, $(\ker f)^\perp \supset \operatorname{im} f^*$ holds because we have $f^*(\beta)(u) = \beta(f(u)) = 0$ for all $u \in \ker f$ with $\beta \in W^*$. Second, $\dim(\ker f)^\perp = \dim U - \dim \ker f = \operatorname{rank} f = \operatorname{rank} f^*$ holds due to the rank–nullity theorem. Then, we obtain $(\ker f)^\perp = \operatorname{im} f^*$. A similar proof is applied for the second statement. Now, we come back to circuit theory and define $B^1 = \operatorname{im} d \subset C^1$. From $(\ker \partial)^\perp = \operatorname{im} d$, we obtain $(Z_1)^\perp = B^1$. This means that there is a potential $\varphi \in C^0$ which satisfies $V = -d\varphi$ for a voltage distribution V.

■ Summary of Circuit Equations

The discussions so far show the duality between KCL and KVL. These results are summarized in Figure 4. Importantly, the degree of freedom for currents and voltages constrained by KCL and KVL is given by $\dim Z_1 + \dim B^1 = \dim C_1$. On the other hand, a circuit element along each edge gives the relation between the current and voltage on the edges, and we have $\dim C_1$ equations with respect to all the circuit elements. Therefore, the current and voltage distributions are unambiguously determined.

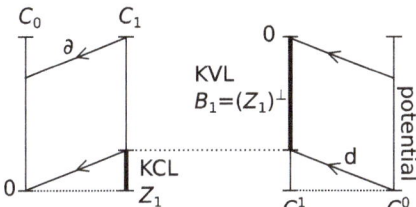

Figure 4. Duality between Kirchhoff's current and voltage laws.

2.2. Planar Graph as Cellular Paving

Consider the series and parallel resistors shown in Figure 5. Series resistors R_1 and R_2 have the composite resistance $R = R_1 + R_2$. On the other hand, parallel resistors R'_1 and R'_2 have the total resistance R', satisfying $1/R' = 1/R'_1 + 1/R'_2$. We can clearly see the duality between resistance and conductance (given by an inverse relationship) for series and parallel resistances. This duality universally holds in more general situations, as we show in Section 2.5. In this subsection, we set up a fundamental geometric structure to establish circuit duality.

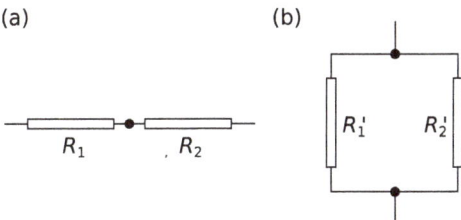

Figure 5. (a) series and (b) parallel resistors with composite resistances R and R', respectively. The duality between resistance and conductance appears as $R = R_1 + R_2$ and $1/R' = 1/R'_1 + /R'_2$.

■ **2-Chains in Planar Graphs**

The graph shown in Figure 1b is *planar*, i.e., its edges intersect only at their nodes. For a planar graph, we can define faces. In Figure 6, we show directed faces $\mathcal{F} = \{f_1, f_2, f_3\}$, where the internal direction of each face is represented by the directed circle. Elements of $\mathcal{F}$ are called 2-cells. Note that we include the unbounded face f_3 outside the circuit. The area of face f_3 can be finite if we consider the planar graph to be on a sphere. The vector space generated by $\mathcal{F}$ is denoted by C_2. It is natural to define a boundary operator $\partial : C_2 \to C_1$, such that $e^j(\partial f_i) = 1$ [$e^j(\partial f_i) = -1$] if $e_j \in \mathcal{E}$ is included in f in the same [opposite] direction; otherwise, $e^j(\partial f_i) = 0$. For Figure 6, we have

$$\partial[f_1\ f_2\ f_3] := [\partial f_1\ \partial f_2\ \partial f_3] = [e_1\ e_2\ e_3\ e_4] \begin{bmatrix} 1 & 0 & 1 \\ 1 & -1 & 0 \\ 0 & 1 & 1 \\ 1 & 0 & 1 \end{bmatrix}. \qquad (4)$$

Note that $\partial \circ \partial = 0$ holds, i.e., the boundary of a cell boundary is empty.

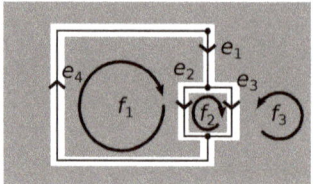

Figure 6. Faces in a planar circuit.

■ Cellular Paving

In the previous examples, cells fully paved the two-dimensional plane without any gap. Such a paving is often called a mesh in finite element analysis. Naturally, a *p-cell* is defined as a *p*-dimensional directed face, which is extended from a 0-cell (point), 1-cell (edge), and 2-cell (face). Cellular paving with the cells can be rigorously formulated in higher-dimensional spaces. The boundary of each element is represented by a combination of lower-dimensional cells. Strictly speaking, a cellular paving of some region R in a manifold is a finite set of open directed p-cells such that (i) two distinct cells do not intersect, (ii) the union of all cells is R, and (iii) if the closures of two cells c and c' meet, their intersection is the closure of a unique cell c'' [62]. However, we only need to grasp the concept of cellular paving with an intuitive sense.

For a cellular paving $\mathcal{K}$ in an m-dimensional region, we can define $C_i(\mathcal{K})$ ($i = 0, 1, \cdots, m$) generated from i-cells with a boundary operator $\partial : C_p(\mathcal{K}) \to C_{p-1}(\mathcal{K})$ satisfying $\partial \circ \partial = 0$. Following a similar treatment for graphs, $Z_p(\mathcal{K}) = \ker\left(\partial : C_p(\mathcal{K}) \to C_{p-1}(\mathcal{K})\right)$ and $B_p(\mathcal{K}) = \text{im}\left(\partial : C_{p+1}(\mathcal{K}) \to C_p(\mathcal{K})\right)$ are defined. Next, we consider a dual space $C^p(\mathcal{K}) = (C_p(\mathcal{K}))^*$. As dual counterparts of $Z_p(\mathcal{K})$ and $B_p(\mathcal{K})$, we can define $Z^p(\mathcal{K}) = \ker\left(d : C^p(\mathcal{K}) \to C^{p+1}(\mathcal{K})\right)$ and $B^p(\mathcal{K}) = \text{im}\left(d : C^{p-1}(\mathcal{K}) \to C^p(\mathcal{K})\right)$. To extract topological characteristics, homology and cohomology groups are defined as $H_p(\mathcal{K}) = Z_p(\mathcal{K})/B_p(\mathcal{K})$ and $H^p(\mathcal{K}) = Z^p(\mathcal{K})/B^p(\mathcal{K})$, respectively. It is known that $\dim H_p(\mathcal{K})$ and $\dim H^p(\mathcal{K})$ do not depend on specific cellular paving, but only on the original m-dimensional region.

2.3. Inner and Outer Orientations

■ Intuitive Explanation of Inner and Outer Orientations

Thus far, cells were assumed to be (internally) oriented. In this subsection, we clarify the definition of orientation and provide a deeper explanation that is crucial for understanding the circuit duality. First, one can intuitively grasp two types of orientation defined for a surface in a three-dimensional space. In Figure 7a, the surface is internally oriented, where "internally" means that the orientation is defined through the internal coordinates of the surface, regardless of whether or not the surface is embedded in the three-dimensional space. On the other hand, we could consider an outer orientation of the surface as the normal-vector field as shown in Figure 7b. Outer orientation refers to the transverse direction of the surface and involves ambient space. To discuss how much fluid traverses the surface, we do not use an inner-oriented surface, but an outer-oriented one. These two types of orientation are strictly defined below and are important to express various physical quantities.

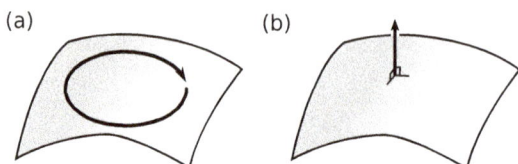

Figure 7. Two types of orientation: (**a**) inner and (**b**) outer orientation of a surface in a *three-dimensional* space.

■ Inner Orientation of Vector Space and Cell

As a starting point, we define an inner orientation of a vector space U. Consider two ordered bases $(e_1, e_2, \cdots, e_m)$ and $(e'_1, e'_2, \cdots, e'_m)$ in U. A basis transformation is represented by $e'_i = \sum_{j=1}^{m} e_j P^j{}_i$. When $\det P > 0$, the direction associated with $(e_1, e_2, \cdots, e_m)$ is considered to be the same as that with $(e'_1, e'_2, \cdots, e'_m)$; otherwise, their directions are different. Then, bases are classified as two different equivalence classes. We can choose one specific class to be positive orientation. If a vector space U has such a positive orientation, U is called inner-oriented. Generally, a tangent space $T_P M$ is a p-dimensional vector space composed of tangent vectors at a point P on a p-dimensional surface (manifold) (Figure 8a). If tangent spaces on a p-cell are continuously oriented, it is called inner-oriented. See a schematic picture shown in Figure 8b.

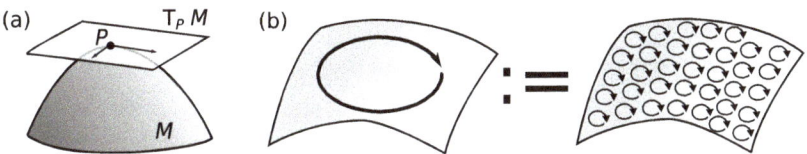

Figure 8. (a) tangent space $T_P M$ at a point P on a surface M; (b) inner-oriented 2-cell with continuously oriented tangent spaces.

■ Outer Orientation of Vector Space and Cell

Next, we introduce an outer orientation. Consider a linear subspace W in a vector space U. When we choose an orientation of the quotient vector space U/W, we say that W is *outer-oriented*. The outer orientation is naturally extended to a p-cell S in an m-dimensional manifold M. Here, M is a total space including S and called ambient space. Because of $T_P S \subset T_P M$, we can consider $T_P M / T_P S$ as a vector space whose (nonzero) elements are considered to be transverse to $T_P S$. If a p-cell has continuous orientation of $T_P M / T_P S$ for all $P \in S$, the cell is called outer-oriented. An example of an outer-oriented 2-cell is shown in Figure 9.

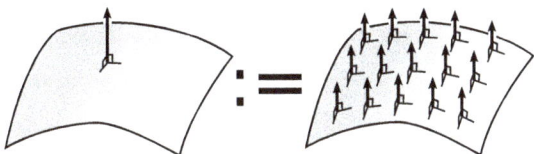

Figure 9. Outer-oriented 2-cell S in a three-dimensional space is continuously outer-oriented in all tangent spaces.

For a planar paving, we summarize all types of cells in Figure 10. An outer-oriented 0-cell is best suited to represent a rotation around a point in the two-dimensional plane. Fluid flow transverse to an edge is also represented by an outer-oriented 1-cell.

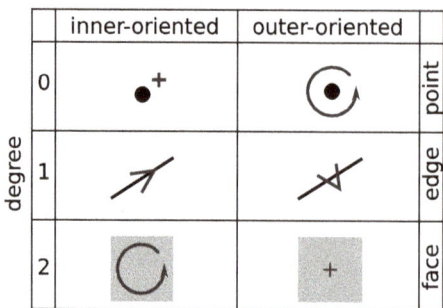

Figure 10. Inner- and outer-oriented cells in a *two-dimensional plane*.

■ Inner-Orientation Representation for Outer Orientation

Outer orientation can be represented by two inner orientations depending on the orientation of ambient space. Let us see this representation in an example in a two-dimensional plane. Consider an outer-oriented edge in a planar graph. With a given orientation of the plane (two-dimensional Euclid space $\mathbb{E}_2$), we can convert the outer orientation into an inner one as shown in Figure 11. The inner orientation is induced by rotating the outer orientation by 90°, where the rotational direction is determined by the ambient-space orientation. Importantly, the inner direction is reversed when we choose the opposite ambient-space orientation. Thus, an outer-oriented cell $\check{s}$ can be represented by two different inner-oriented cells as $\check{s} = \{\check{s}_o | o = \circlearrowleft, \circlearrowright\}$ with $\check{s}_\circlearrowleft = -\check{s}_\circlearrowright$. The above intuitive discussion can be generalized in other dimensional spaces as shown in Appendix A.

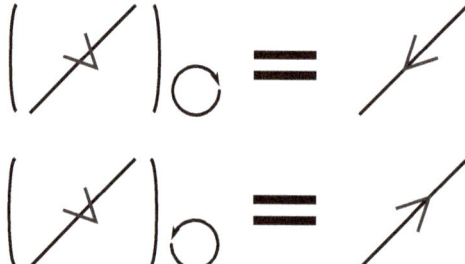

Figure 11. Inner-oriented components of an outer-oriented edge in a two-dimensional plane. The plane orientations are given by the subscripts ($\circlearrowleft, \circlearrowright$).

■ Outer-Orientation Representation for Inner Orientation

The previous discussions on orientations are based on representations of an outer-oriented cell by inner-oriented cells. We would like to remark on a dual perspective: an inner-oriented cell can be represented by outer-oriented cells. This perspective is indicated in Figure 12. To obtain an outer orientation, the inner orientation of the edge is rotated by 90° in the *reverse* direction of a given orientation of the plane.

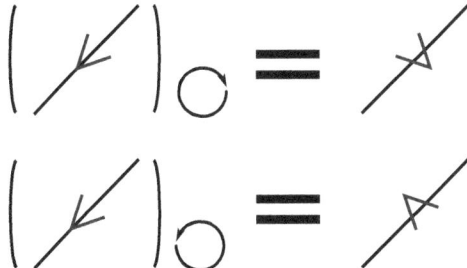

Figure 12. Outer-oriented components of an inner-oriented edge in a two-dimensional plane.

■ Operation for Outer-Oriented Chains and Cochains

Next, consider a cellular paving $\check{\mathcal{K}}$ with outer-oriented cells. We can define a vector space $C_p(\check{\mathcal{K}})$ generated from p-cells in $\check{\mathcal{K}}$. Using the representation of an outer-oriented cell by inner-oriented cells, we can define $\partial : C_p(\check{\mathcal{K}}) \to C_{p-1}(\check{\mathcal{K}})$ so that ∂ individually acts for each inner-oriented component. For example, an inner-oriented component of the boundary of an outer-oriented 1-cell in a plane is depicted in Figure 13. The boundary operations for outer-oriented cells in a two-dimensional plane are summarized in Figure 14.

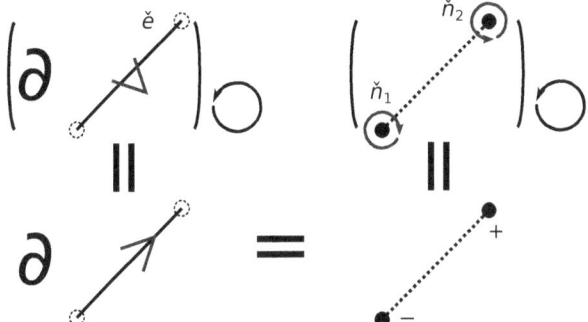

Figure 13. Boundary operation for an outer-oriented 1-cell in a two-dimensional plane is defined through the inner-oriented component.

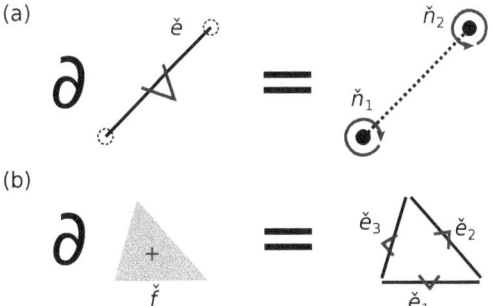

Figure 14. Boundary operation for an outer-oriented (**a**) edge and (**b**) face in a two-dimensional plane.

As a dual counterpart of $C_p(\check{\mathcal{K}})$, the dual space $C^p(\check{\mathcal{K}}) = (C_p(\check{\mathcal{K}}))^*$ is defined by applying the general theory for linear spaces. An outer-oriented cochain in $C^p(\check{\mathcal{K}})$ is also represented by two inner-oriented cochains and the coboundary operator d for outer-oriented cochains is naturally defined.

2.4. Essence of Poincaré Duality

■ Dual Paving

For a cellular paving $\mathcal{K}$ with inner-oriented cells, we can compose a dual cellular paving $\mathcal{K}^\star$ with outer-oriented cells. Here, we focus on ambient space of a two-dimensional plane to introduce the concept. As shown in Figure 15a, each face f_i in $\mathcal{K}$ is converted to an outer-oriented point $\check{n}_i$ in $\mathcal{K}^\star$. Each edge e_j in $\mathcal{K}$ is also replaced with an outer-oriented edge $\check{e}_j$ in $\mathcal{K}^\star$ transverse to the original edge. The obtained edge is connected to a point in $\mathcal{K}^\star$, if the original edge is included in the original face. Furthermore, each point n_j in $\mathcal{K}$ is converted to a face $\check{f}_j$ in $\mathcal{K}^\star$ as shown in Figure 15b. The face is adjacent to an edge in $\mathcal{K}^\star$, if the original point is connected to the original edge. Under this composition, the orientation of each cell is naturally inherited from the original cell to the dual cell.

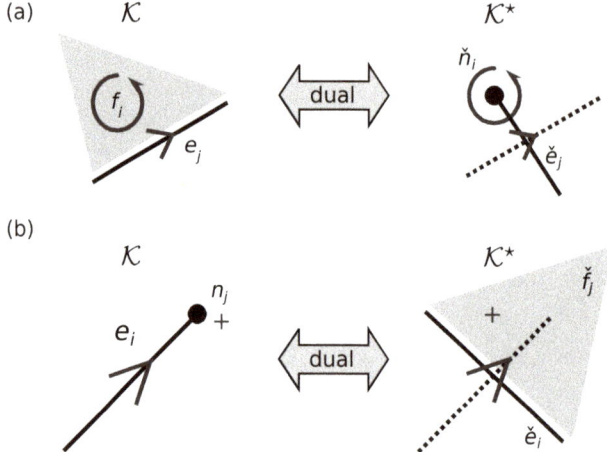

Figure 15. Correspondence between a cellular paving $\mathcal{K}$ and its dual paving $\mathcal{K}^\star$. (**a**) a face f_i in $\mathcal{K} \leftrightarrow$ a point $\check{n}_i$ in $\mathcal{K}^\star$; (**b**) a point n_j in $\mathcal{K} \leftrightarrow$ a face $\check{f}_j$ in $\mathcal{K}^\star$. In both figures, an edge e_j in $\mathcal{K}$ corresponds to an edge $\check{e}_j$ in $\mathcal{K}^\star$.

Let us explicitly see this composition in the previous example, where the original cellular paving is shown again in Figure 16a. By applying the above procedure for $\mathcal{K}$ shown in Figure 16a, we obtain the dual paving $\mathcal{K}^\star$ as shown in Figure 16b.

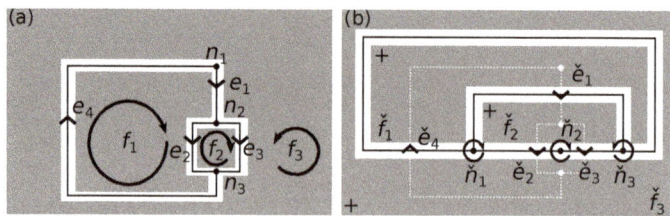

Figure 16. (**a**) cellular paving and (**b**) its dual paving.

Poincaré Duality

For this dual paving, we have the following matrix representations of ∂:

$$\partial[\check{f}_1\ \check{f}_2\ \check{f}_3] = [\check{e}_1\ \check{e}_2\ \check{e}_3\ \check{e}_4] \begin{bmatrix} 1 & -1 & 0 \\ 0 & 1 & -1 \\ 0 & 1 & -1 \\ -1 & 0 & 1 \end{bmatrix}, \tag{5}$$

$$\partial[\check{e}_1\ \check{e}_2\ \check{e}_3\ \check{e}_4] = [\check{n}_1\ \check{n}_2\ \check{n}_3] \begin{bmatrix} 1 & 1 & 0 & 1 \\ 0 & -1 & 1 & 0 \\ 1 & 0 & 1 & 1 \end{bmatrix}. \tag{6}$$

Comparing Equations (5) and (6) with Equations (1) and (4), we have

$$\left[\partial : C_1(\mathcal{K}) \to C_0(\mathcal{K})\right] = -\left[\partial : C_2(\mathcal{K}^\star) \to C_1(\mathcal{K}^\star)\right]^\mathrm{T}, \tag{7}$$

$$\left[\partial : C_2(\mathcal{K}) \to C_1(\mathcal{K})\right] = \left[\partial : C_1(\mathcal{K}^\star) \to C_0(\mathcal{K}^\star)\right]^\mathrm{T}, \tag{8}$$

where the square brackets indicate matrix representation. The above relations $n^j(\partial e_i) = -\check{e}^i(\partial \check{f}_j)$ and $e^j(\partial f_i) = \check{n}^i(\partial \check{e}_j)$ hold from Figure 15. These relations universally hold even in a higher-dimensional space. Recalling $[d : C^p(\mathcal{K}) \to C^{p+1}(\mathcal{K})] = [\partial : C_{p+1}(\mathcal{K}) \to C_p(\mathcal{K})]^\mathrm{T}$, we obtain the following commutative diagram:

$$\begin{array}{ccccc} C_2(\mathcal{K}) & \xrightarrow{\partial} & C_1(\mathcal{K}) & \xrightarrow{\partial} & C_0(\mathcal{K}) \\ \downarrow{\star_2} & & \downarrow{\star_1} & & \downarrow{\star_0} \\ C^0(\mathcal{K}^\star) & \xrightarrow{d} & C^1(\mathcal{K}^\star) & \xrightarrow{-d} & C^2(\mathcal{K}^\star) \end{array} \tag{9}$$

Here, we have isomorphisms $\star_p : C_p(\mathcal{K}) \to C^{m-p}(\mathcal{K}^\star)$ with $m = 2$, where $\star_0 : n_i \mapsto \check{f}^i$, $\star_1 : e_i \mapsto \check{e}^i$, and $\star_2 : f_i \mapsto \check{n}^i$. The dual counterpart of Equation (9) is given by

$$\begin{array}{ccccc} C^2(\mathcal{K}) & \xleftarrow{d} & C^1(\mathcal{K}) & \xleftarrow{d} & C^0(\mathcal{K}) \\ \uparrow{\star^2} & & \uparrow{\star^1} & & \uparrow{\star^0} \\ C_0(\mathcal{K}^\star) & \xleftarrow{\partial} & C_1(\mathcal{K}^\star) & \xleftarrow{-\partial} & C_2(\mathcal{K}^\star) \end{array} \tag{10}$$

with $\star^i = (\star_i)^*$ using the dual map $(\star_i)^*$ of $\star_i$. These correspondences naturally hold in higher-dimensional spaces and lead to Poincaré duality: $H_p(M) \cong H^{m-p}(M)$ for a compact orientable m-dimensional manifold M [63].

2.5. Dual Circuits

Poincaré Duality and Circuit Duality

Now, let us introduce dual circuits. Consider a circuit on a planar graph. The planar graph is seen as a cellular paving $\mathcal{K}$ for a two-dimensional plane (or a sphere surface). Because there is no nontrivial loop on a plane or sphere, we have the following equation:

$$\mathrm{im}\left(\partial : C_2(\mathcal{K}) \to C_1(\mathcal{K})\right) = \ker\left(\partial : C_1(\mathcal{K}) \to C_0(\mathcal{K})\right). \tag{11}$$

Let $I \in Z_1(\mathcal{K})$ and $V \in B^1(\mathcal{K})$ be current and voltage distributions satisfying KCL and KVL, respectively. We set a reference resistance $R_{\mathrm{ref}}(= 1/G_{\mathrm{ref}})$ to exchange a current and voltage. Consider current $I^\star := G_{\mathrm{ref}}(\star^1)^{-1}(V)$ and voltage $V^\star := R_{\mathrm{ref}} \star_1 (I)$ distributions for a circuit on $\mathcal{K}^\star$, where $\star_1 : e_i \mapsto \check{e}^i$ and $(\star^1)^{-1} : e^i \mapsto \check{e}_i$ are defined with $\star^1 := (\star_1)^*$. Note that $I^\star$ and $V^\star$ are outer-oriented, but we can obtain an inner-oriented component for a given orientation of the plane. By combining Equation (9) with Equation (11), KCL and KVL are shown to hold for $I^\star$ and $V^\star$ as explained below. Here, we check KVL for $V^\star$. From Equation (11), a current distribution $I \in Z_1(\mathcal{K})$ can be written as $I = \partial F$ with "face" (or mesh) currents $F \in C_2(\mathcal{K})$. Then, we have $V^\star = R_{\text{ref}} \text{d}(\star_2(F))$, which indicates $V^\star \in B^1(\mathcal{K}^\star)$. A similar discussion to KCL holds for $I^\star$.

A current and voltage relation along a circuit element located at each edge $e_i \in \mathcal{E}$ is written by $h_i(I, V) = 0$ for $I \in Z_1(\mathcal{K})$, and $V \in B^1(\mathcal{K})$. As a dual relation, we define $h_i^\star(I^\star, V^\star) := h_i(G_{\text{ref}}(\star_1)^{-1}(V^\star), R_{\text{ref}} \star^1 (I^\star))$. If we assign a circuit element with $h_i^\star(I^\star, V^\star) = 0$ for each edge $\check{e}_i$ in $\mathcal{K}^\star$, $I^\star$ and $V^\star$ give a solution of the circuit on $\mathcal{K}^\star$. For example, consider a resistance with Ohm's law $V = IR$ with scalar V, I, and R. Substituting $V \to I^\star R_{\text{ref}}$, $I \to G_{\text{ref}} V^\star$, we have $I^\star = G^\star V^\star$ with $G^\star = R/(R_{\text{ref}})^2$. Importantly, we obtain

$$RR^\star = (R_{\text{ref}})^2 \tag{12}$$

for $R^\star = 1/G^\star$. The derived circuit on $\mathcal{K}^\star$ is called a *dual circuit*. The dual relation is summarized in Table 1.

Table 1. Dual relations in electrical circuits.

Circuit	Dual Circuit
Current distribution I	Voltage distribution $V^\star = R_{\text{ref}} \star_1 (I)$
Voltage distribution V	Current distribution $I^\star = G_{\text{ref}}(\star^1)^{-1}(V)$
Face current F	Potential $\varphi^\star = -R_{\text{ref}} \star_2 (F)$
Potential φ	Face current $F^\star = G_{\text{ref}}(\star^0)^{-1}(\varphi)$
Voltage source V_s	Current source $I_s^\star = G_{\text{ref}} V_s$
Current source I_s	Voltage source $V_s^\star = R_{\text{ref}} I_s$
Resistance R	Conductance $G^\star = (G_{\text{ref}})^2 R$
Conductance G	Resistance $R^\star = (R_{\text{ref}})^2 G$

For the previous example shown in Figure 17a, we construct a dual circuit as shown in Figure 17b. Here, we have $I_s^\star = V_S/R_{\text{ref}}$, $R_i^\star = (R_{\text{ref}})^2/R_i$, and take a specific orientation (↻) of the plane to obtain an inner orientation of the current source. We can clearly see the duality between series and parallel connections in Figure 17. Therefore, the concept of dual circuits is considered as a generalization of series–parallel duality.

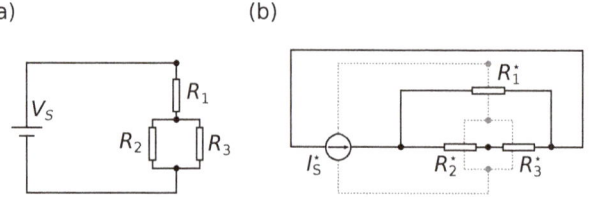

Figure 17. (a) example circuit and (b) its dual circuit.

■ Duality for Composite Resistances

Consider a one-port network N composed of resistors. A voltage source is attached to N as shown in Figure 18a. The network N is characterized by an equivalent resistance $R = V_s/I_s$, where V_s and I_s are the voltage and current along the source, respectively. Consider the dual counterpart as shown in Figure 18b. The dual network $N^\star$ is characterized by $R^\star = V_s^\star/I_s^\star$ with the current $I_s^\star$ and voltage $V_s^\star$ along the source. From the corresponding duality, we have

$$RR^\star = (R_{\text{ref}})^2, \tag{13}$$

where R_{ref} is the reference resistance.

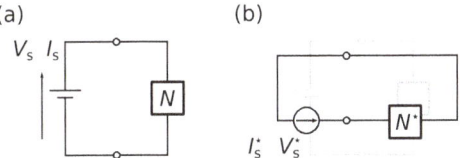

Figure 18. (a) one-port network with a voltage source and (b) its dual counterpart.

2.6. Self-Dual Circuit

If the *self-dual* relation $N = N^*$ holds for a one-port network N, the composite resistance satisfies $R = R^*$. In this case, we obtain $R = R^* = R_{\text{ref}}$ without solving circuit equations. As an example, we consider a bridge circuit shown in Figure 19a and its dual counterpart over R_{ref} is given in Figure 19b. The self-dual condition is written as $R_1 R_2 = (R_{\text{ref}})^2$. Under the self-dual condition, the circuit behaves as an effective resistor with R_{ref}.

Thus far, we only consider circuits with resistors, but the extension to alternating-current (AC) circuits is obvious. Under this extension, the real-number field ($\mathbb{R}$) is replaced with the complex one ($\mathbb{C}$). In AC circuits, resistances are replaced with complex impedances. Consider the example of an AC circuit shown in Figure 19c. This circuit is obtained from Figure 19a by replacing $R_1 \to Z_1 = j\omega L$, $R_2 \to Z_2 = 1/(j\omega C)$, where ω is an angular frequency. The self-dual condition is given by $R_{\text{ref}} = \sqrt{L/C}$. Under the self-duality, the circuit surprisingly behaves as a frequency-independent resistor with R_{ref}, although capacitors and inductors exhibit frequency-dependent response. This result can be interpreted to mean that the frequency dependency of capacitors and inductors negate each other when the whole system is self-dual. Therefore, self-dual circuits provide a powerful way to produce constant-resistance circuits [64].

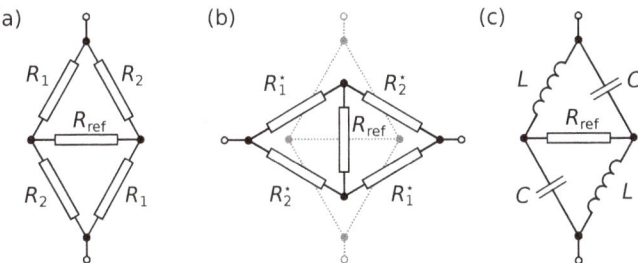

Figure 19. (a) bridge circuit and (b) its dual circuit; (c) alternating-current bridge circuit.

3. Zero Backscattering from Self-Duality

Signals can propagate in a uniform transmission line without backscattering. In this section, we associate self-duality with the zero-backscattering condition. Furthermore, we show that a large phase shift without backscattering in Huygens' metasurfaces can be understood by using a self-dual circuit model.

3.1. Self-Dual Transmission Lines

In this section, we consider signal propagation in a transmission line, which can be expressed by an LC ladder network [65]. In contrast to previous research on duality in transmission lines [66], we provide a circuit theoretical interpretation of the characteristic impedance of a transmission line in terms of self-dual response described in Section 2.

■ Telegraph Equations

A coaxial cable, which is composed of an inner conductor as a signal line and an outer conductor as a ground shield, forms capacitors between two conductors and inductors with a magnetic field around the inner conductor as shown in Figure 20a. With inductance $\bar{L}$ and capacitance $\bar{C}$ per unit length, the system is discretized as the LC ladder composed of $L_{i+\frac{1}{2}} = \bar{L}\Delta z$ and $C_i = \bar{C}\Delta z$ in a unit length Δz as shown in Figure 20b. In the i-th segment, Kirchhoff's voltage and current laws are, respectively, expressed as

$$L_{i-\frac{1}{2}} \frac{dI_{i-\frac{1}{2}}}{dt} = V_{i-1} - V_i, \tag{14}$$

$$C_i \frac{dV_i}{dt} = I_{i-\frac{1}{2}} - I_{i+\frac{1}{2}}. \tag{15}$$

In the continuum limit of $\Delta z \to 0$, the above equations become

$$\bar{L}\frac{\partial I}{\partial t} = -\frac{\partial V}{\partial z}, \tag{16}$$

$$\bar{C}\frac{\partial V}{\partial t} = -\frac{\partial I}{\partial z}, \tag{17}$$

which are well known as the telegraph equations. Generally, $\bar{L}(z)$ and $\bar{C}(z)$ can depend on z.

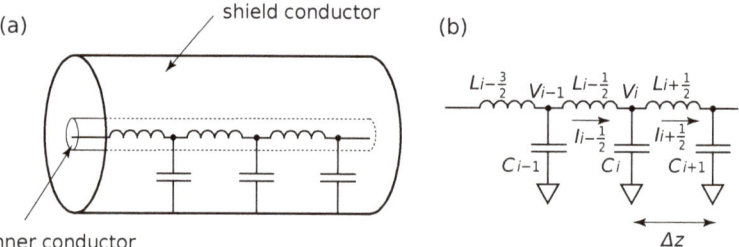

Figure 20. (a) coaxial cable and (b) its circuit model.

■ Zero Backscattering in Self-Dual Transmission Lines

Here, we see that the self-dual transmission line does not cause backscattering. If a transmission line has a constant impedance $Z = \sqrt{\bar{L}(z)/\bar{C}(z)}$ independent of z, then the transmission line becomes self-dual: the telegraph equations with a uniform Z are invariant under duality transformation $(V, I) \leftrightarrow (R_{\text{ref}}I, V/R_{\text{ref}})$ with $R_{\text{ref}} = Z$. In terms of the new variables $\eta_\pm = V \pm ZI$, the self-dual telegraph equations are written as

$$\frac{\partial \eta_\pm}{\partial z} \pm \frac{1}{v_0}\frac{\partial \eta_\pm}{\partial t} = 0, \tag{18}$$

where $v_0(z) = 1/\sqrt{\bar{L}\bar{C}}$. With arbitrary functions $f_\pm$, the general expression of the solution can be provided by $\eta_\pm(t,z) = f_\pm(t \mp \int^z dz'/v_0(z'))$, which corresponds to a propagating wave with velocity $\pm v_0(z)$ without backscatterng. If $Z(z)$ depends on z, self-duality is broken and backscattering may be observed.

■ Circuit-Theoretical Derivation of Zero Backscattering

Next, we give a circuit-theoretical derivation of zero backscattering for a self-dual system. A transmission line composed of LC ladder circuits is shown in Figure 21a, where inductances

$L_{i+\frac{1}{2}}$ and capacitances C_i may depend on the position i. The dual circuit with respect to a global resistance R_{ref} is illustrated as shown in Figure 21b. The transformed circuit elements are given as

$$L_i^\star = C_i (R_{\text{ref}})^2, \quad C_{i+\frac{1}{2}}^\star = L_{i+\frac{1}{2}} / (R_{\text{ref}})^2. \tag{19}$$

In the limit of small $L_i^\star$ and $C_{i+\frac{1}{2}}^\star$, each capacitance can be shifted to next position because the potential difference across each inductance is negligibly small. By repeating the shift of the capacitances $C_i \to C_{i+1}$ in the original circuit as shown in Figure 21c, we obtain the same circuit as Figure 21b ($L_i^\star = L_{i+\frac{1}{2}}$ and $C_{i+\frac{1}{2}}^\star = C_i$) under the following condition for any i:

$$R_{\text{ref}} = \sqrt{\frac{L_{i+\frac{1}{2}}}{C_i}}. \tag{20}$$

In other words, the LC ladder network with constant $L_{i+\frac{1}{2}}/C_i$ is self-dual for impedance inversion with respect to R_{ref} defined by Equation (20). In this way, the *global* parameter R_{ref} is linked with the *local* impedance $Z_i = \sqrt{L_{i+\frac{1}{2}}/C_i}$, which is defined by $L_{i+\frac{1}{2}}$ and C_i in each site.

Next, we consider excitation of the self-dual LC ladder by a voltage source V connected at the left-hand side as shown in Figure 21d. The dual circuit is illustrated in Figure 21e, where the current source is given by $I_s = V_s / R_{\text{ref}}$. By repeating the shift of the capacitances $C_i \to C_{i+1}$ as depicted in Figure 21f, Figure 21e can be identified with Figure 21f. As discussed in Section 2.6, the self-dual circuit is characterized by the effective resistance R_{ref} as shown in Figure 21g. Imagine that we connect a uniform transmission line (T1) with a characteristic impedance of R_{ref} to the half-infinite circuit (T2). When the signal propagates from T1 to T2, any backscattering does not appear due to the impedance matching.

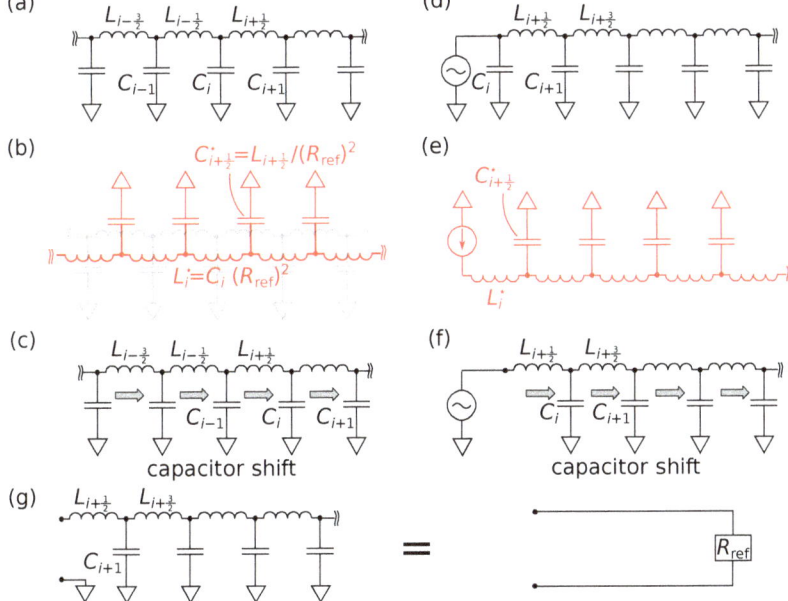

Figure 21. (**a**) LC ladder circuit and (**b**) its dual circuit; (**c**) capacitors in the circuit (**a**) are shifted; (**d**) LC ladder circuit excited by a voltage source and (**e**) its dual circuit; (**f**) capacitors in the circuit (**d**) are shifted; (**g**) input impedance of the LC ladder circuit.

■ Heaviside Condition

Self-duality can be realized in transmission lines with inductance $L_{i+\frac{1}{2}} = \bar{L}\Delta z$, capacitance $C_i = \bar{C}\Delta z$, conductance $G_i = \bar{G}\Delta z$, and resistance $R_{i+\frac{1}{2}} = \bar{R}\Delta z$ as shown in Figure 22. The self-dual condition is given by

$$\sqrt{\frac{\bar{L}}{\bar{C}}} = \sqrt{\frac{\bar{R}}{\bar{G}}}. \tag{21}$$

This self-dual condition is nothing but a no-distortion signal transmission condition derived by Heaviside [67]. Due to the self-duality, frequency-independent response is realized and backscattering vanishes, while a signal decays as it propagates.

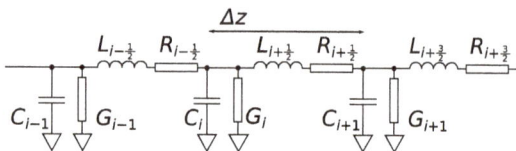

Figure 22. LC ladder with resistance and conductance.

3.2. Circuit Model for Huygens' Metasurface

Two-dimensional artificial structures called metasurfaces have been extensively investigated for controlling the amplitude and phase of transmitted and/or reflected electromagnetic waves [68]. The wavefront control of light can be realized by designing metasurfaces with spatial variations of phase responses [69–72]. The amplitude and phase responses of metasurfaces are generally interdependent. Nevertheless, it is possible to control the phase of the transmitted light with constant power transmission, which could be 100% in ideal conditions without losses, by carefully designing the resonant components of the metasurfaces. Metasurfaces for the arbitrary control of transmission properties, or amplitude and phase control, are called Huygens' metasurfaces, which have been introduced by Pfeiffer and Grbic [73]. In this subsection, we clarify the role of self-duality in Huygens' metasurfaces.

■ Transmission and Reflection for Huygens' Metasurfaces

It is assumed that monochromatic plane electromagnetic waves with a specific polarization are normally incident on an isotropic metasurface placed at $z = 0$ in a vacuum with a wave impedance of Z_0. In this paper, the variable with a tilde represents the complex amplitude of a harmonically oscillating quantity $A = \tilde{A}e^{j\omega t} + \text{c.c.}$, where c.c. denotes the complex conjugate of the preceding term. The complex amplitudes of macroscopic electric and magnetic fields, which are averaged over the typical scale of metasurface elements, are represented by $\tilde{E}_-$ and $\tilde{H}_-$ ($\tilde{E}_+$ and $\tilde{H}_+$) for the input (output) side $z \leq 0$ ($z \geq 0$) in the proximity to the surface. Although electric and magnetic fields are represented by vectors, we here focus on scalar amplitudes for a specific polarization. Electromagnetic response of the metasurface is characterized by two parameters: electric sheet admittance Y_e and magnetic sheet impedance Z_m. The averaged electric fields $\tilde{E}_{av} = (\tilde{E}_- + \tilde{E}_+)/2$ induce surface currents $\tilde{K} = Y_e\tilde{E}_{av}$, which demand the boundary condition $\tilde{H}_- - \tilde{H}_+ = \tilde{K}$ on $z = 0$ [74]. In the same way, the magnetic counterpart can be considered, and the averaged magnetic fields $\tilde{H}_{av} = (\tilde{H}_- + \tilde{H}_+)/2$ produce surface magnetic currents $\tilde{K}_m = Z_m\tilde{H}_{av}$, which require the boundary condition $\tilde{E}_- - \tilde{E}_+ = \tilde{K}_m$. The boundary conditions are summarized as

$$\tilde{H}_- - \tilde{H}_+ = Y_e \frac{\tilde{E}_- + \tilde{E}_+}{2}, \tag{22}$$

$$\tilde{E}_- - \tilde{E}_+ = Z_m \frac{\tilde{H}_- + \tilde{H}_+}{2}. \tag{23}$$

For the incident wave propagating in the $+z$ direction with the electric field $\tilde{E}_{in} e^{-jkz}$ and the magnetic field $\tilde{H}_{in} e^{-jkz} (= \tilde{E}_{in} e^{-jkz}/Z_0)$, the total electric and magnetic fields $(\tilde{E}, \tilde{H})$ are represented as $(\tilde{E}_{in} e^{-jkz} + \varrho \tilde{E}_{in} e^{jkz}, \tilde{H}_{in} e^{-jkz} - \varrho \tilde{H}_{in} e^{jkz})$ in $z \leq 0$ and $(\tau \tilde{E}_{in} e^{-jkz}, \tau \tilde{H}_{in} e^{-jkz})$ in $z \geq 0$, where τ and ϱ are the amplitude transmission and reflection coefficients. By substituting these fields at the metasurface ($z = 0$) into Equations (22) and (23), the amplitude transmission and reflection coefficients are obtained as

$$\tau = \frac{(Z_2 - Z_1)Z_0}{(Z_0 + Z_1)(Z_0 + Z_2)},$$
$$\varrho = \frac{Z_1 Z_2 - (Z_0)^2}{(Z_0 + Z_1)(Z_0 + Z_2)},$$
(24)

where $Z_1 = Z_m/2$ and $Z_2 = 2/Y_e$. Hence, the reflection vanishes for $Z_1 Z_2 = (Z_0)^2$. In addition to the no-reflection condition, if Z_1 and Z_2 are purely imaginary impedances, which are expressed as $Z_1 = (Z_0)^2/Z_2 = jbZ_0$ with a dimensionless number $b \in \mathbb{R}$, the transmission coefficient can be written as

$$\tau = \frac{1 - jb}{1 + jb}.$$
(25)

The incident waves are perfectly transmitted through the metasurface due to the fact that $|\tau|^2 = 1$ for any b, and the transmitted waves acquire a phase of $\theta = -2 \arctan b$. Such metasurfaces, which are typical examples of Huygens' metasurfaces, realize arbitrary phase shift θ without losses in an ideal case by tailoring the design of the metasurface structures. Both of the electric and magnetic responses are indispensable for the no-reflection condition.

■ Circuit Model for Huygens' Metasurfaces

The propagation of electromagnetic waves in a vacuum can be modeled as signal propagation in a transmission line with the wave impedance Z_0, and the metasurface is represented by circuit elements inserted in the transmission line as shown in Figure 23a. In this model, the electric field $\tilde{E}$ and magnetic field $\tilde{H}$ are replaced with voltage $\tilde{V}$ and current $\tilde{I}$, respectively; therefore, the circuit model of the metasurface should satisfy the following conditions:

$$\tilde{I}_- - \tilde{I}_+ = Y_e \frac{\tilde{V}_- + \tilde{V}_+}{2},$$
(26)

$$\tilde{V}_- - \tilde{V}_+ = Z_m \frac{\tilde{I}_- + \tilde{I}_+}{2}.$$
(27)

A circuit called a lattice circuit as shown in Figure 23b satisfies the above conditions [75]. The electric and magnetic responses are represented by impedances $Z_2 (= 2/Y_e)$ and $Z_1 (= Z_m/2)$, respectively. Equations (26) and (27) can be confirmed separately for Figure 23b, considering excitation by waves from both sides of the metasurface. For in-phase excitation $\tilde{V}_- = \tilde{V}_+$, all currents are sunk into the bridge circuit, and the currents become antiphase $\tilde{I}_- = -\tilde{I}_+$. There is no voltage across Z_1, and the currents flow only in Z_2. Hence, we obtain $\tilde{V}_- = Z_2 \tilde{I}_-$, which is identical to Equation (26) for $\tilde{V}_- = \tilde{V}_+$ and $\tilde{I}_- = -\tilde{I}_+$. In the opposite case, $\tilde{V}_- = -\tilde{V}_+$ and $\tilde{I}_- = \tilde{I}_+$, the currents flow only in Z_1, and $\tilde{V}_- = Z_1 \tilde{I}_-$, so the equation that corresponds to Equation (27) can be derived.

■ Zero Backscattering Due to Self-Duality

For Figure 23a, $Z_{in} = \tilde{V}_-/\tilde{I}_-$ provides the input impedance for the metasurface, or lattice circuit, followed by the transmission line in $z > 0$. The uniform semi-infinite transmission line in $z > 0$ can be regarded as a resistor with an impedance of Z_0 as shown in Figure 21g. As a result, the total system viewed from the input side $z < 0$ is well described by a bridge circuit as shown in Figure 23c, and Z_{in} is identical to the impedance of the bridge circuit. As described in Section 2.6, the bridge circuit satisfying $Z_1 Z_2 = (Z_0)^2$ is self-dual for the inversion center Z_0, and the input impedance Z_{in} is always Z_0. The reflection vanishes under this condition, where the wave impedance Z_0 is impedance-matched to the load represented by the bridge circuit. As a result, all energy is transmitted to the transmission line in $z > 0$. Thus, the no-reflection condition for Huygens' metasurfaces is interpreted in terms of self-duality.

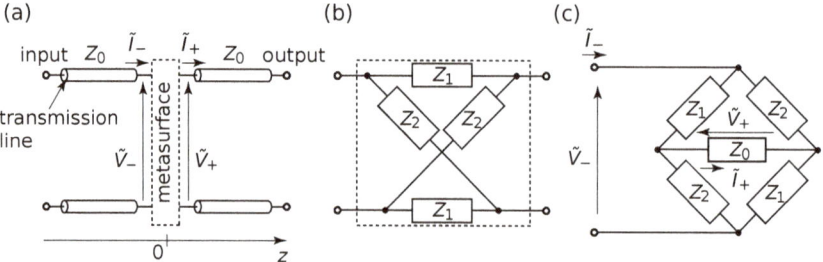

Figure 23. (a) circuit model for propagating electromagnetic waves incident on a metasurface; (b) circuit model of Huygens' metasurfaces with $Z_1 = Z_m/2$ and $Z_2 = 2/Y_e$; (c) lumped circuit model for the metasurface followed by a semi-infinite transmission line.

4. Keller–Dykhne Duality

Circuit duality can be extended for a continuous system. As with circuits, the effective response of a continuous system is associated with that of the dual one; thus, a self-dual response is automatically guaranteed. Such a constraint can induce critical behaviors of self-dual systems. Here, these topics are reviewed. Next, we introduce differential forms to clearly extract the structure of the duality in a continuous system. The correspondence between the dualities in continuous and discrete systems is formulated through discretization of continuous fields. In this section, we always use the right-hand vector product, in other words, A, B, and $A \times B$ obey the right-hand rule.

4.1. Two-Dimensional Resistive Sheets

Consider an electric field $E(x, y)$ and current density $K(x, y)$ on a two-dimensional resistive sheet located at $z = 0$ with a sheet conductance $G(x, y)$. Here, we assume that the fields inside the thin sheet are uniform along z and omit z-dependency for the fields. Thus, we treat the fields as two-dimensional vector fields independent of z. An example configuration is shown in Figure 24a. The physical dimensions of E, K, and G are V/m, A/m, and $1/\Omega$, respectively. From KVL and KCL, we obtain

$$\nabla \times E = 0, \tag{28}$$
$$\nabla \cdot K = 0. \tag{29}$$

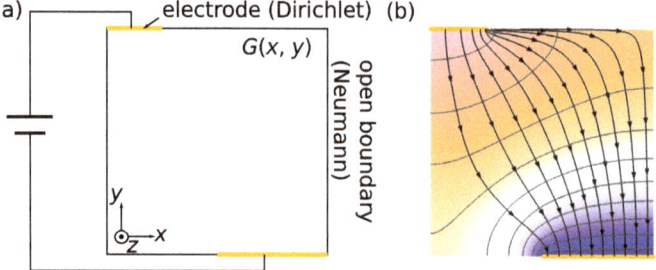

Figure 24. (a) two-dimensional resistive sheet with boundary conditions; (b) solution for a sheet with a constant sheet conductance $G(x, y) = G_{\text{ref}}$. Black lines represent the current flow. The potential is shown as a color map with isopotential gray contours.

Note that these equations can be directly obtained from Maxwell's equations; by omitting the time-derivative terms for steady states, we can obtain Equations (28) and (29) from Faraday's law and the law of charge conservation. Ohm's law is given by

$$K = GE, \tag{30}$$

where G is a conductance and generally a 2×2 matrix. For a metallic electrode, the boundary condition is written as

$$n \times E = 0, \tag{31}$$

where n is the unit vector normal to the boundary. For an open boundary, the boundary condition is given by

$$n \cdot K = 0. \tag{32}$$

From Equation (28), the electric field is represented by

$$E = -\nabla \varphi \tag{33}$$

with a potential $\varphi(x, y)$. Combining Equation (33) with Equations (29) and (30), we obtain

$$\nabla \cdot G \nabla \varphi = 0. \tag{34}$$

The boundary of the i-th electrode is specified by the Dirichlet boundary condition

$$\varphi = \varphi_i, \tag{35}$$

which is constant along the boundary. The open boundary is given by the Neumann boundary condition

$$n \cdot G \nabla \varphi = 0. \tag{36}$$

For a simplified system with a constant scalar conductance, we have the Laplace equation

$$\nabla^2 \varphi = 0 \tag{37}$$

from Equation (34). An example solution of the Laplace equation calculated by COMSOL Multiphysics® is shown in Figure 24b.

4.2. Duality in Laplace Equation

In this subsection, we discuss duality in Laplace equations [76]. The solution of the Laplace equation is called a *harmonic function*. A harmonic function can be considered as a part of a holomorphic function. To see this fact, consider a holomorphic function $w(z) = u(z) + jv(z)$ with $z = x + jy$, $u(z) = \text{Re}[w(z)]$, and $v(z) = \text{Im}[w(z)]$. The holomorphism leads to the Cauchy–Riemann equations:

$$\frac{\partial u}{\partial x} = \frac{\partial v}{\partial y}, \tag{38}$$

$$\frac{\partial u}{\partial y} = -\frac{\partial v}{\partial x}. \tag{39}$$

These equations can be expressed as

$$\begin{bmatrix} \frac{\partial v}{\partial x} \\ \frac{\partial v}{\partial y} \end{bmatrix} = J \begin{bmatrix} \frac{\partial u}{\partial x} \\ \frac{\partial u}{\partial y} \end{bmatrix} \tag{40}$$

with $J = \begin{bmatrix} 0 & -1 \\ 1 & 0 \end{bmatrix}$, which induces counterclockwise 90° rotation. As we differentiate Equations (38) and (39) along x and y, respectively, and combine the results, we obtain

$$\nabla^2 u = 0, \tag{41}$$

which states that u is harmonic. Similarly, v is also harmonic. Here, v is called a *harmonic conjugate* of u.

If u is given, how can we obtain its harmonic conjugate v? Focusing on $v(x,y) = v(x_0, y_0) + \int_{\text{path}:(x_0,y_0)\to(x,y)} (\nabla v) \cdot d\mathbf{r}$ with Equation (40), we have

$$v(x,y) = v(x_0, y_0) + \int_{\text{path}:(x_0,y_0)\to(x,y)} (J\nabla u) \cdot d\mathbf{r}, \tag{42}$$

where $\mathbf{r} = [x\ y]^T$ and (x_0, y_0) is a fixed point. We have assumed that the considered region is simply connected to define Equation (42). When we consider a small displacement $\Delta \mathbf{r}$ along a line $v(x,y) = \text{const.}$, we have $(J\nabla u) \cdot \Delta \mathbf{r} = 0$, which leads to $\nabla u \parallel \Delta \mathbf{r}$. Therefore, u and v constitute an orthogonal coordinate around the point $\nabla u \neq 0$, as shown in Figure 25a. In this subsection, the operation of taking the harmonic conjugate is treated as a duality transformation.

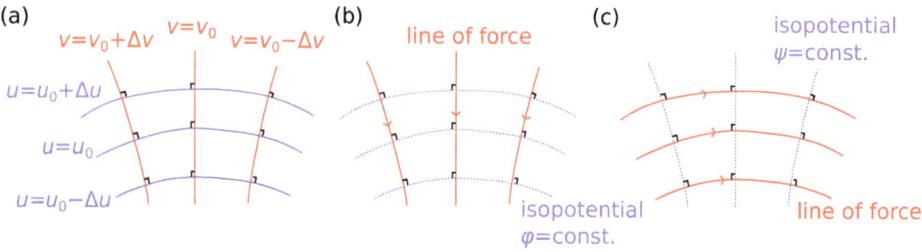

Figure 25. (a) holomorphic function $w(z) = u(z) + jv(z)$ defines an orthogonal coordinate around a point with $dw/dz \neq 0$; (b) harmonic potential φ and the lines of force $-\nabla \varphi$; (c) harmonic conjugate ψ for φ and the lines of force $-\nabla \psi$.

The above result induces duality for the potential problem of the Laplace equation. Let ψ be the harmonic conjugate of a harmonic potential φ. The relation between φ and ψ is depicted in Figure 25b,c. Now, we come back to resistive sheet problems. The current stream lines and isopotential contours are replaced with each other under the harmonic conjugate as shown in the simplest example of Figure 26. Furthermore, we can see that the harmonic conjugate interchanges the Dirichlet and the Neumann boundary conditions because of the 90° rotation.

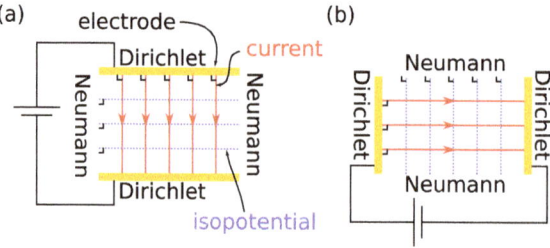

Figure 26. Current and potential distributions for (**a**) original and (**b**) its dual resistive sheets with a uniform conductance.

4.3. Generalized Duality

Harmonic duality can be extended to a two-dimensional resistive sheet with a spatially inhomogeneous sheet conductance $G(x,y)$. Keller proved the duality for a system composed of two different conductances [6], and Dykhne generalized it to a system with an arbitrary scalar function $G(x,y)$ [7]. The extended duality is often called *Keller–Dykhne duality*. Furthermore, Mendelson generalized the duality for a tensor $G(x,y)$ [8].

Here, we review the derivation of the duality. As with circuit duality, we set a reference resistance $R_{ref}(=1/G_{ref})$. Referring to a 90° rotation in Equation (40), we introduce the following dual fields on $z=0$:

$$E^\star = R_{ref} J K = R_{ref} e_z \times K, \tag{43}$$

$$K^\star = G_{ref} J E = G_{ref} e_z \times E, \tag{44}$$

with the unit vector e_z along the z direction. The 90° rotation interchanges divergence and rotation of a two-dimensional vector field $v(x,y)$ as

$$\nabla \times (e_z \times v) = e_z \nabla \cdot v, \tag{45}$$

$$\nabla \cdot (e_z \times v) = -e_z \cdot (\nabla \times v), \tag{46}$$

where we used $\nabla \times (A \times B) = A(\nabla \cdot B) - B(\nabla \cdot A) + (B \cdot \nabla)A - (A \cdot \nabla)B$ and $\nabla \cdot (A \times B) = B \cdot (\nabla \times A) - A \cdot (\nabla \times B)$. Thus, KVL and KCL for $E^\star$ and $K^\star$ automatically follow: $\nabla \times E^\star = 0$ and $\nabla \cdot K^\star = 0$. In addition, Ohm's law in Equation (30) is converted to

$$K^\star = G^\star E^\star \tag{47}$$

with

$$G^\star = (G_{ref})^2 J G^{-1} J^{-1}. \tag{48}$$

For $R = G^{-1}$ and $R^\star = (G^\star)^{-1}$, we obtain

$$R^\star J R J^{-1} = (R_{ref})^2. \tag{49}$$

Therefore, $E^\star$ and $K^\star$ give a solution for the sheet with $G^\star$ under interchanging the Dirichlet and Neumann boundary conditions. Note that G can be a tensor. For a scalar G, we simply obtain $G^\star G = (G_{ref})^2$.

4.4. Effective Response and Duality

Duality can relate the effective response of an original sheet with its dual counterpart. To see this statement, consider a resistive sheet with two terminals as shown in Figure 27a. Between electrodes, we have the voltage $V = -\int_{C_1} E \cdot dr$ and current flow $I = \int_{C_2} K \cdot n_2 \, dr$. The effective conductance between the terminals is defined as $G_{eff} = I/V$. The dual system is also shown in Figure 27b, and the voltage and current are represented by $V^\star = -\int_{C_2} E^\star \cdot dr$ and $I^\star = \int_{C_1} K^\star \cdot n_1 dr$, respectively. Using $A \cdot (B \times C) = B \cdot (C \times A) = C \cdot (A \times B)$, we obtain

$$V^\star = R_{ref} I, \tag{50}$$

$$I^\star = G_{ref} V. \tag{51}$$

Therefore, the effective conductance $G_{eff}^\star$ for the dual system is given by

$$G_{eff}^\star G_{eff} = (G_{ref})^2. \tag{52}$$

If the system is self-dual and passive, which means that there is no active element with a negative resistance, then $G_{\text{eff}} = G_{\text{eff}}^\star = G_{\text{ref}}$ automatically follow. Thus, self-duality determines the effective response regardless of the structural geometry.

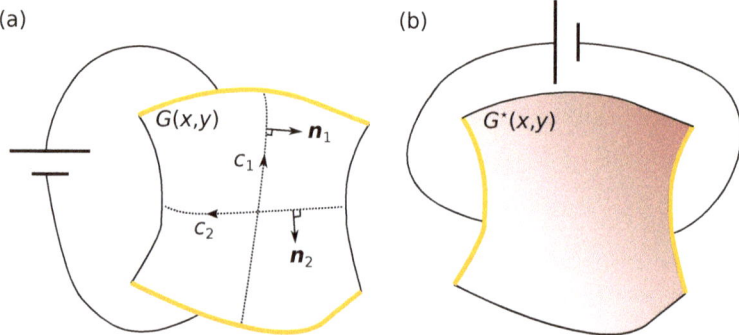

Figure 27. (a) resistive sheet with a sheet conductance $G(x,y)$ and two terminals; (b) corresponding counterpart with $G^\star = (G_{\text{ref}})^2 JG^{-1}J^{-1}$. Unit normal vectors are denoted by n_1 and n_2 for curves c_1 and c_2, respectively.

4.5. Self-Duality and Singularity

The self-dual effective response can sometimes predict a critical behavior of the system. To see how critical behavior appears in self-dual systems, we consider an ideal checkerboard with different admittances Y_1 and Y_2, as shown in Figure 28a. Here, the AC response with an angular frequency ω is discussed for the checkerboard. For a capacitive admittance $Y_1 = j\omega C$ and an inductive admittance $Y_2 = (j\omega L)^{-1}$, the checkerboard is self-dual with respect to a reference conductance $G_{\text{ref}} = \sqrt{Y_1 Y_2} = \sqrt{C/L}$. Then, we obtain a real effective admittance $Y_{\text{eff}} = G_{\text{ref}}$ due to the self-duality. The positive real admittance indicates that the system is lossy. However, the system is lossless because the effective admittance composed only of capacitors and inductors must be purely imaginary. Therefore, we do not have a physical solution for such a checkerboard composed of capacitors and inductors. Thus, the lossless ideal checkerboard is *singular*.

The above observation can be also interpreted from a branch cut for the self-dual admittance $Y_{\text{eff}}(Y_1, Y_2) = \sqrt{Y_1 Y_2}$ [77]. We fix Y_2 at a point of the negative imaginary axis and gradually displace Y_1 from Y_2 as shown in Figure 28b. When we consider $Y_{\text{eff}}(Y_1) = \sqrt{Y_1 Y_2}$ as a function of Y_1, $Y_{\text{eff}}(Y_1)$ must have a branch cut in the complex plane. The previous discussion clearly shows that the branch cut is located along the positive imaginary axis. This result indicates that two approaches from $\text{Re}(Y_1) > 0$ and $\text{Re}(Y_1) < 0$ regions to a point at the singular branch lead to different values of Y_{eff}. The effective admittance on the branch is critically sensitive to loss or gain of the system.

The above result can be generalized for arbitrary Y_2. Using linearity

$$Y_{\text{eff}}(sY_1, sY_2) = sY_{\text{eff}}(Y_1, Y_2) \tag{53}$$

for a scalar s, we can write Y_{eff} as $Y_{\text{eff}} = Y_2 R_{\text{ref}} Y_{\text{eff}}(\eta G_{\text{ref}}, G_{\text{ref}})$ with $\eta = Y_1/Y_2$. Therefore, all system characteristics are derived from $y_{\text{eff}}(\eta) := R_{\text{ref}} Y_{\text{eff}}(\eta G_{\text{ref}}, G_{\text{ref}})$ satisfying $y_{\text{eff}}(\eta) = \eta\, y_{\text{eff}}(\eta^{-1})$ due to the self-duality $Y_{\text{eff}}(Y_1, Y_2) = Y_{\text{eff}}(Y_2, Y_1)$. Because singular Y_1 is represented by $Y_1 = sY_2\,(s < 0)$ in the previous discussion, self-dual $y_{\text{eff}}(\eta)$ has a branch cut along the negative real axis of η.

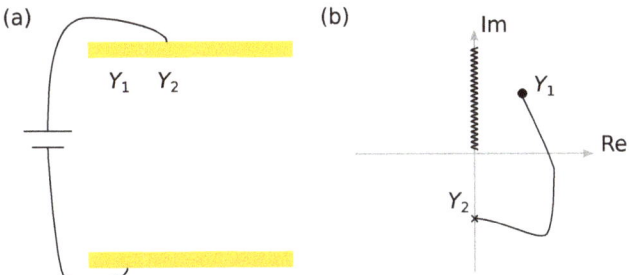

Figure 28. (a) ideal checkerboard sheet with sheet admittances Y_1 and Y_2; (b) domain of definition for the effective admittance $Y_{\text{eff}}(Y_1) = \sqrt{Y_1 Y_2}$. The branch cut along the positive imaginary axis is indicated by a wavy line.

4.6. Differential-Form Approach for Duality

Although Keller–Dykhne duality was formulated through vector analysis, the essence of the duality can be vividly extracted by differential forms. Furthermore, differential forms are suitable to discretize continuous fields to circuit systems. Discretizing differential forms, we bridge between Keller–Dykhne duality and circuit duality.

Here, let us introduce differential forms and the exterior derivative in plain terms. To this end, we only focus on Cartesian coordinates. For more technical details of differential forms, see [59,78].

■ Covector and 1-Form

First, consider an electric field at a point (x_0, y_0). The Cartesian basis is denoted by $\{e_x, e_y\}$. A displacement $\Delta r = \Delta x e_x + \Delta y e_y$ and the electric potential difference $\Delta \varphi$ are related through

$$-\Delta \varphi = E_x(x_0, y_0) \Delta x + E_y(x_0, y_0) \Delta y. \tag{54}$$

It is possible to consider an electric field E as a linear function as $E : \Delta r \mapsto -\Delta \varphi$. Physically, we may understand the electric field as an apparatus that measures the (minus) electric potential difference for a displacement Δr. Such a linear map from a vector to a scalar is called a *covector*. Introducing a dual basis $\{dx, dy\}$ with respect to the Cartesian basis $\{e_x, e_y\}$, we can write the covector as

$$E(x_0, y_0) = E_x(x_0, y_0) dx + E_y(x_0, y_0) dy. \tag{55}$$

The action of the *interior product* between $E(x_0, y_0)$ and a displacement vector Δr is written as

$$\Delta r \lrcorner E(x_0, y_0) = E_x(x_0, y_0) \Delta x + E_y(x_0, y_0) \Delta y. \tag{56}$$

From the definition, we have $e_x \lrcorner dx = 1$, $e_y \lrcorner dx = 0$, $e_y \lrcorner dy = 1$, and $e_x \lrcorner dy = 0$. A covector α should be depicted as a set of parallel lines with an outer orientation as shown in Figure 29a. As the covector becomes stronger, the lines become denser. The interior product $v \lrcorner \alpha$ gives the signed number of lines that the vector v pierces. Note that we consider a limit operation in the strict sense as follows: we set a small resolution value ε and define a set of lines $\{L_m | m \in \mathbb{Z}\}$ with $L_m = \{r | r \lrcorner \alpha = m\varepsilon\}$. The number of the lines of $\{L_m\}$ pierced by v is denoted as $N(\varepsilon)$. Then, we obtain $v \lrcorner \alpha = \lim_{\varepsilon \to 0} \varepsilon N(\varepsilon)$.

The electric field is given by a covector field

$$E(x, y) = E_x(x, y) dx + E_y(x, y) dy, \tag{57}$$

which smoothly depends on (x, y). Covector fields are generally called *1-forms*. An example of a 1-form is shown in Figure 29b. For a general 1-form α, $v \lrcorner \alpha$ is evaluated at the tangent space to which

v belongs: $v \lrcorner \alpha = v \lrcorner \alpha(x_0, y_0)$ for v whose starting point is (x_0, y_0). The 1-form α can be integrated along a curve c as

$$\int_c \alpha := \lim_{N \to \infty} \sum_{i=0}^{N-1} \Delta r_i \lrcorner \alpha, \qquad (58)$$

where curve $c(t) : [0,1] \to \mathbb{E}_2$ is discretized as $\Delta r_i = c(t_{i+1}) - c(t_i)$ with $t_i = i/N$ ($\mathbb{E}_2$: two-dimensional Euclidean space). Note that Δr_i is considered to be in the tangent space of $c(t_i)$. This definition can be linearly extended for any 1-chain c, and a 1-form is considered as a 1-cochain. By integrating, we can express a voltage difference as $V = -\int_c E$.

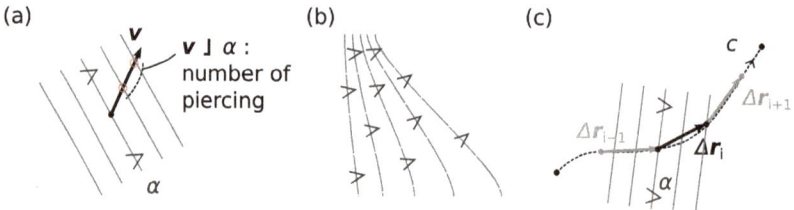

Figure 29. (a) illustration of a covector α. The signed number of lines that a vector v pierces is given by $v \lrcorner \alpha$. The positive direction of α is depicted by carets; (b) covector field (1-form); (c) discretized curve c with displaced vectors Δr_i to define the integral of a 1-form α along c.

■ Tensors and Products

Similar to a covector, a (covariant) tensor maps p input vectors to a scalar output: $T(v_1, v_2, \cdots, v_p)$, which has linearity in each slot. The interior product of a vector v and a tensor T can be defined as $v \lrcorner T = T(v, \sqcup, \cdots)$, which indicates that the first slot is filled with v and the other slots are left waiting for inputs ($\sqcup$).

To obtain a higher-order tensor, other products are introduced. First, the tensor product of two covectors α and β is defined as follows: $\alpha \otimes \beta(u, v) = \alpha(u)\beta(v)$ for u, v in a vector space U. For 1-forms, the tensor product operates on each tangent space.

Second, the wedge product for two 1-forms α and β is defined as an antisymmetrized tensor

$$\alpha \wedge \beta = \frac{1}{2}(\alpha \otimes \beta - \beta \otimes \alpha). \qquad (59)$$

The wedge product of two 1-forms satisfies

$$\alpha \wedge \beta = -\beta \wedge \alpha, \qquad (60)$$

$$\alpha \wedge \alpha = 0. \qquad (61)$$

As $dx \wedge dx = dy \wedge dy = 0$, only $dx \wedge dy$ plays an important role. For $u = u^x e_x + u^y e_y$ and $v = v^x e_x + v^y e_y$, we have

$$dx \wedge dy(u, v) = \frac{1}{2} \det \begin{bmatrix} u^x & v^x \\ u^y & v^y \end{bmatrix}. \qquad (62)$$

Due to the antisymmetry $dx \wedge dy(u, v) = -dx \wedge dy(v, u)$, $dx \wedge dy$ measures a signed area of a triangle spanned by u and v. In Figure 30a, we illustrate $dx \wedge dy$ as directed circles and $dx \wedge dy(u, v)$ counts the number of circles inside the triangle spanned by u and v. If the direction from u to v is the same (opposite) as the direction of the circles, the circles are counted as positive (negative) numbers.

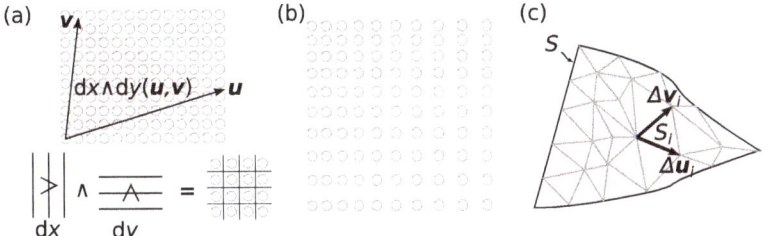

Figure 30. (a) $dx \wedge dy$ as directed circles; (b) example of a 2-form; (c) discretization for integration.

■ 2-Forms and Integration

We can also consider a field $\omega = f(x,y)dx \wedge dy$ with a scalar function $f(x,y)$, where ω is called a 2-form. An example of a 2-form is shown in Figure 30b, where the dense circle area has a stronger field than the sparse area. For integration, we discretize a directed 2-cell S as $S = \bigcup_i S_i$ with disjoint small triangles S_i (Figure 30c). A 2-form can be integrated on S as

$$\int_S \omega = \lim \sum_i f(x_i, y_i) \omega(\Delta u_i, \Delta v_i), \tag{63}$$

where Δu_i and Δv_i span a triangle S_i with the same direction of S and $(x_i, y_i) \in S_i$. The limit is taken for finer meshes. Generally, S can be an arbitrary 2-chain. Then, a 2-form can be considered as a 2-cochain. Note that integration of a general tensor T without antisymmetry cannot be defined and the antisymmetry is essential for integration. To define the integral, the integration over subdivided triangles should be the same as that of the original triangle. At least, $T(u,v) = T\left(u, \frac{u+v}{2}\right) + T\left(\frac{u+v}{2}, v\right)$ is required for arbitrary u and v. Considering $u = v$, we obtain $T(u,u) = 0$ for arbitrary u. Then, the antisymmetry $T(u,v) = -T(v,u)$ must hold due to $T(u+v, u+v) = 0$.

■ Exterior Derivative

The exterior derivative for a scalar function $f(x,y)$ is defined as

$$df = \frac{\partial f}{\partial x}dx + \frac{\partial f}{\partial y}dy, \tag{64}$$

which corresponds to the gradient of the function f. For a 1-form $\alpha = \alpha_x dx + \alpha_y dy$, we define the exterior derivative as

$$d\alpha = (d\alpha_x) \wedge dx + (d\alpha_y) \wedge dy. \tag{65}$$

By direct calculation,

$$d\alpha = \left(\frac{\partial \alpha_y}{\partial x} - \frac{\partial \alpha_x}{\partial y}\right) dx \wedge dy,$$

this corresponds to a rotation. By using the exterior derivative, KVL is represented as $dE = 0$.

■ Stokes' Theorem

How may one relate an exterior derivative to the boundary of a chain? Green's theorem

$$\oint_{\partial S}(f dx + g dy) = \iint_S \left(\frac{\partial g}{\partial x} - \frac{\partial f}{\partial y}\right) dx dy \tag{66}$$

can be rewritten as the following equation for a 2-chain S with a 1-form α:

$$\int_S d\alpha = \int_{\partial S} \alpha. \tag{67}$$

Equation (67) generally holds for higher-dimensional spaces, and it is known as Stokes' theorem. Stokes' theorem relates the boundary operator ∂ with the exterior derivative d.

■ **Twisted 1-Form**

Next, we consider a representation of current density. As discussed in Section 2.3, current flow should be calculated for an outer-oriented 1-chain, so the current density is an outer-oriented 1-cochain. The outer-oriented 1-chain $\check{c}$ was represented as $\check{c} = \{\check{c}_o | o = \circlearrowright, \circlearrowleft\}$ by using the inner-oriented 1-chain $\check{c}_o$ depending on a plane orientation o. Thus, current density K should be represented as $K = \{K_o | o = \circlearrowright, \circlearrowleft\}$ with the two 1-forms satisfying

$$K_{-o} = -K_o, \tag{68}$$

where $-o$ represents the opposite orientation of o. The set of two 1-forms satisfying Equation (68) is called a *twisted* 1-form. On the other hand, ordinary forms are called *untwisted*. A twisted 1-form $\check{\alpha}$ at a point P is an outer-oriented covector $\check{\alpha}_P$, which is depicted as the inner-oriented lines in Figure 31a. To stress the twist, we put the check symbol ($\check{\ }$) for twisted objects, but sometimes omit the mark to reduce the notation complexity. In Figure 31b, the twisted 1-form K is depicted by (local) stream lines. We can count the total current flow across a curve by the integration. The integration of K along an outer-oriented 1-chain $\check{c}$ is defined as

$$\int_{\check{c}} K := \int_{\check{c}_o} K_o. \tag{69}$$

The exterior derivative of K is also defined as

$$(dK)_o = dK_o. \tag{70}$$

KCL states that

$$\int_{\partial \check{S}} K = \int_{\check{S}} dK = 0 \tag{71}$$

for any outer-oriented 2-cell $\check{S}$. Considering small $\check{S}$, we obtain $dK = 0$ as KCL.

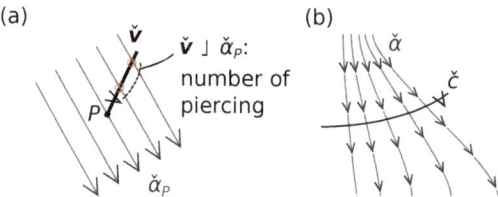

Figure 31. (a) interior product between an outer-oriented vector $\check{v}$ and an outer-oriented covector $\check{\alpha}_P$ at a point P; (b) integration of a twisted 1-form $\check{\alpha}$ along an outer-oriented 1-chain $\check{c}$.

■ **Metric Tensor**

To express Ohm's law, we will define the Hodge star operation. To define the Hodge star operation, we need to introduce a metric tensor. In the two-dimensional Euclidean space $\mathbb{E}_2$, we can define the inner product of vectors $u = u^x e_x + u^y e_y$ and $v = v^x e_x + v^y e_y$ as

$$g(u, v) = u \cdot v \tag{72}$$

with $u \cdot v = u^x v^x + u^y v^y$. Here, g is called the *metric tensor*. We can define a covector $g^{\flat}(u)$ for a vector u satisfying

$$g^{\flat}(u) = g(u, \sqcup). \tag{73}$$

Therefore, $v \lrcorner g^\flat(u) = g(u,v)$ holds. The operation of $g^\flat$ on e_x and e_y is graphically shown in Figure 32. The inverse map of $g^\flat$ is written as $g^\sharp = (g^\flat)^{-1}$.

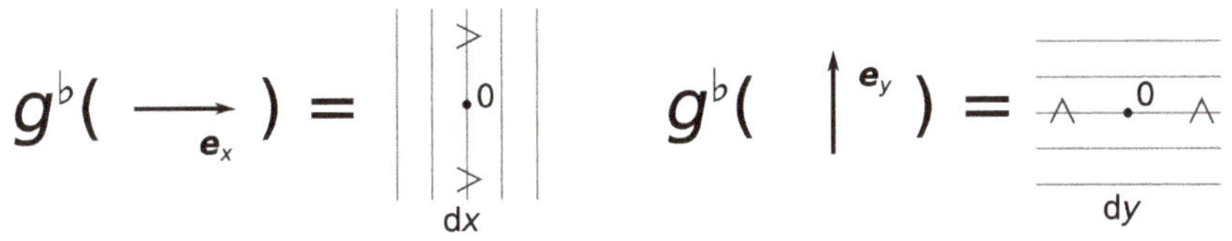

$$g^\flat(\xrightarrow{e_x}) = \begin{array}{c} \uparrow y \\ \bullet 0 \\ \downarrow \\ dx \end{array} \qquad g^\flat(\uparrow^{e_y}) = \begin{array}{c} \\ \overline{\wedge \quad \bullet 0 \quad \wedge} \\ dy \end{array}$$

Figure 32. Conversion between a vector and a covector.

■ Area Form as a Twisted 2-Form

With respect to the metric of $\mathbb{E}_2$, $\{e_x, e_y\}$ is an orthogonal basis: $g(e_i, e_j) = \delta_{ij}$ (δ_{ij}: Kronecker delta). Using the orthogonal coordinate, we can define a twisted 2-form "Area" as

$$\text{Area}_\circlearrowleft = dx \wedge dy, \tag{74}$$
$$\text{Area}_\circlearrowright = -dx \wedge dy. \tag{75}$$

The area form measures the *unsigned* area of an outer-oriented 2-chain.

■ Hodge Star

Now, we define the Hodge star operation for a 1-form α as

$$(\star \alpha)_o = g^\sharp(\alpha) \lrcorner \text{Area}_o \tag{76}$$

with respect to an orientation o of the plane. The star operator $\star$ maps a 1-form to a twisted 1-form. Naturally, we can define a Hodge star operation for a twisted 1-form $\breve{\alpha}$ as:

$$\star \breve{\alpha} = g^\sharp(\breve{\alpha}_o) \lrcorner \text{Area}_o. \tag{77}$$

Then, the Hodge operator maps a twisted 1-form to an untwisted 1-form. The multiple operations of $\star$ are shown in Figure 33. In this figure, we can see

$$\star \star = -\text{Id}, \tag{78}$$

where Id is the identity operator. Therefore, $\star$ defines the complex structure in the two-dimensional plane. By using the Hodge star, we can represent Ohm's law with a scalar sheet conductance G as $K = G \star E$. Note that the Hodge operator can be defined for other p-forms, but the sign of $\star\star$ generally depends on the order p, the dimension of the space, and the metric signature, rather than Equation (78) [59].

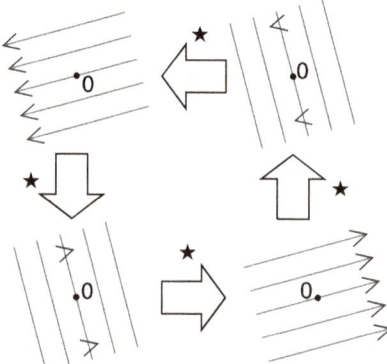

Figure 33. Complex structure of Hodge operations for untwisted and twisted 1-forms.

4.7. Summary of Basic Equations in Differential-Form Approach

Now, we summarize the basic equations with differential forms for a two-dimensional resistive sheet with a nonuniform scalar conductance. The electric field is represented by a 1-form E, while the current density field is given by a twisted 1-form K. KVL and KCL are formulated as

$$dE = 0, \qquad (79)$$
$$dK = 0, \qquad (80)$$

respectively. The scalar Ohm's law is rewritten as

$$K = G \star E \qquad (81)$$

with a scalar sheet conductance $G(= 1/R)$. These equations are schematically shown in Figure 34. Although we only focused on Cartesian coordinates, Equations (79)–(81) are coordinate free. Therefore, we can use an arbitrary coordinate for analysis. Another feature of the differential-form formalism is the exclusion of the metric in Equations (79) and (80). The metric appears through the Hodge star in Equation (81). Thus, Equations (79) and (80) are metric-free equations and easy to be discretized while keeping the geometrical structure, as we see later.

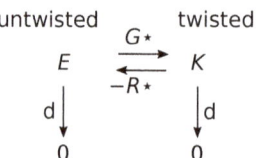

Figure 34. Structure of basic equations in a two-dimensional resistive sheet.

4.8. Keller–Dykhne Duality with Differential Forms

Now, we formulate Keller–Dykhne duality with differential forms. Electric and current fields are represented by untwisted and twisted 1-forms, respectively. To exchange these fields with two different kinds of orientations, we need to fix an orientation ω of the plane. Here, we define a twisted scalar $\Omega^\omega = \{(\Omega^\omega)_o | o = \circlearrowright, \circlearrowleft\}$ satisfying $(\Omega^\omega)_\omega = +1$ and $(\Omega^\omega)_{-\omega} = -1$. The pseudoscalar Ω^ω is regarded as the plane orientation ϖ. For a twisted form $\check{\omega}$, Ω^ϖ extracts the component $\Omega^\varpi \check{\omega} = \check{\omega}_\varpi$. Now, we consider the replacement as

$$E^\star = R_{\text{ref}} \Omega^\varpi K, \tag{82}$$

$$K^\star = -G_{\text{ref}} \Omega^\varpi E \tag{83}$$

with respect to a reference resistance $R_{\text{ref}}(=1/G_{\text{ref}})$. Clearly, these fields satisfy

$$dE^\star = 0, \tag{84}$$

$$dK^\star = 0. \tag{85}$$

From Equation (81) with Equations (82) and (83), we obtain

$$K^\star = G^\star \star E^\star, \tag{86}$$

where we use Equation (78) and $G^\star = (G_{\text{ref}})^2/G$ for a scalar G.

4.9. Discretization

Discretizing a two-dimensional sheet, we can obtain circuit duality again. In this subsection, we rigorously confirm this statement.

■ Discretization

For a sheet region U, we consider a cellular paving $\mathcal{K}$ with nodes $\mathcal{N}$, edges $\mathcal{E}$, and faces $\mathcal{F}$. We write the relations among nodes, edges, and faces as

$$\partial e_i = \sum_{j=1}^{|\mathcal{N}|} n_j \Delta^j{}_i \quad (i = 1, 2, \cdots, |\mathcal{E}|), \tag{87}$$

$$\partial f_i = \sum_{j=1}^{|\mathcal{E}|} e_j \Pi^j{}_i \quad (i = 1, 2, \cdots, |\mathcal{F}|) \tag{88}$$

for $e_i \in \mathcal{E}$ and $f_i \in \mathcal{F}$. The totality of p-forms on U is denoted by $C^p(U)$. We can define $\phi^p : C^p(U) \to C^p(\mathcal{K})$ as

$$[\phi^p(\omega)](c) = \int_c \omega \tag{89}$$

for $\omega \in C^p(U)$ and all $c \in C_p(\mathcal{K})$. Here, ϕ^p makes a continuous field ω discretized. Now, we obtain a commutative diagram:

$$\begin{array}{ccc} C^0(U) \xrightarrow{d} C^1(U) \xrightarrow{d} C^2(U) \\ \downarrow \phi^0 \quad \downarrow \phi^1 \quad \downarrow \phi^2 \\ C^0(\mathcal{K}) \xrightarrow{d} C^1(\mathcal{K}) \xrightarrow{d} C^2(\mathcal{K}) \end{array} \tag{90}$$

Commutativity can be checked as follows. We can calculate for all $f \in C^0(U), c \in C_1(\mathcal{K})$ as

$$[\phi^1(df)](c) = \int_c df = \int_{\partial c} f = [\phi^0(f)](\partial c) = [d\phi^0(f)](c), \tag{91}$$

where we used Stokes' theorem. Similarly, we obtain $\phi^2(d\alpha) = d[\phi^1(\alpha)]$, for all $\alpha \in C^1(U)$. The commutative diagram of Equation (90) indicates that the discretization by ϕ keeps the algebraic structure of d.

Using ϕ, we can discretize an electric field (untwisted 1-form) E as $V = \sum_{i=1}^{|\mathcal{E}|} V_i e^i$ with $V_i = \int_{e_i} E$. Moreover, KVL ($dE = 0$) is discretized as

$$dV = 0, \tag{92}$$

which is explicitly expressed as $\sum_{j=1}^{|\mathcal{E}|} V_j \Pi^j{}_i = 0$. Therefore, $\Pi^j{}_i$ represents the discretized rotation of the field.

For a current density K (twisted 1-form), we should consider the integration on the dual lattice. The set of all twisted p-forms on U is represented as $\check{C}^p(U)$. We have

$$[\check{\phi}^p(\check{\omega})](\check{c}) = \int_{\check{c}} \check{\omega} \tag{93}$$

for $\check{\omega} \in \check{C}^p(U)$ and all $\check{c} \in C_p(\mathcal{K}^\star)$. Then, another diagram similar to Equation (90) is obtained:

$$\begin{array}{ccccc}
\check{C}^0(U) & \xrightarrow{d} & \check{C}^1(U) & \xrightarrow{d} & \check{C}^2(U) \\
\downarrow \check{\phi}^0 & & \downarrow \check{\phi}^1 & & \downarrow \check{\phi}^2 \\
C^0(\mathcal{K}^\star) & \xrightarrow{d} & C^1(\mathcal{K}^\star) & \xrightarrow{d} & C^2(\mathcal{K}^\star)
\end{array} \tag{94}$$

Remembering $(\star_1)^{-1} : C^1(\mathcal{K}^\star) \to C_1(\mathcal{K})$, we can define $I = (\star_1)^{-1}(\check{\phi}^1(K)) = \sum_{i=1}^{|\mathcal{E}|} I^i e_i$ with $I^i = \int_{\check{e}_i} K = \int_{(\star_1)^{-1} e^i} K$. Now, the discretized KCL is obtained as

$$\partial I = 0, \tag{95}$$

for which component representation is $\sum_{j=1}^{|\mathcal{E}|} \Delta^i{}_j I^j = 0$. Therefore, $\Delta^i{}_j$ indicates the discretized minus divergence.

For the discretization of Ohm's law, we interpolate E from E_i as

$$E \approx \sum_{i=1}^{|\mathcal{E}|} E_i w_{e_i}, \tag{96}$$

where $\{w_{e_i}\}$ are called interpolation forms. As the interpolation forms $\{w_{e_i}\}$, so-called Whitney forms can be used [62,79–82]. Now, Ohm's law is discretized as

$$I^i = \sum_{j=1}^{|\mathcal{E}|} \left(\int_{\check{e}_i} G \star w_{e_j} \right) E_j. \tag{97}$$

■ Correspondence between Keller–Dykhne Duality and Circuit Duality

Now, we establish correspondence between Keller–Dykhne duality and circuit duality. These two dualities are related through the diagram shown in Figure 35. In this section, we prove the commutativity of the two paths (1) and (2) in the figure.

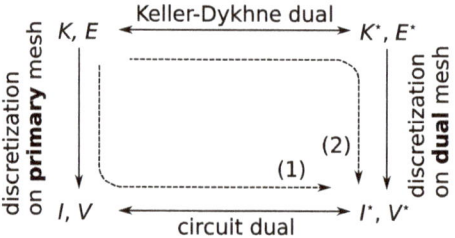

Figure 35. Relation between Keller–Dykhne duality and circuit duality.

(1) For I and V discretized from K and E, we can consider the dual circuit with current $I^\star := G_{\text{ref}}(\star^1)^{-1}(V)$ and voltage $V^\star := R_{\text{ref}} \star_1 (I)$ distributions for a circuit on $\mathcal{K}^\star$:

$$I^\star = G_{\text{ref}} \sum_{i=1}^{|\mathcal{E}|} V_i(\star^1)^{-1}(e^i) = G_{\text{ref}} \sum_{i=1}^{|\mathcal{E}|} V_i \check{e}_i, \tag{98}$$

$$V^\star = R_{\text{ref}} \sum_{i=1}^{|\mathcal{E}|} I^i \star_1 (e_i) = R_{\text{ref}} \sum_{i=1}^{|\mathcal{E}|} I^i \check{e}^i. \tag{99}$$

(2) On the other hand, we discretize $E^\star = R_{\text{ref}} K_\omega$ and $K^\star = -G_{\text{ref}} \Omega^\omega E$ in a dual mesh $\mathcal{K}^\star$. We need to choose a specific orientation ω of the plane, and $(\check{e}_i)_\omega$ is regarded as an inner-oriented edge in $\mathcal{K}^\star$. Here, we introduce the $\star$-conjugate operation to give an outer-oriented dual edge as $e_i{}^\star = \check{e}_i$. The dual edge $(\check{e}_i)_\omega{}^\star$ is outer-oriented, and represented as $((\check{e}_i)_\omega{}^\star)_\omega = -e_i$ (Figure 36), which reflects the complex algebraic structure of the plane. Then, discretized $E^\star$ and $K^\star$ are given as

$$\left(\int_{(\check{e}_i)_\omega} E^\star \right) (\check{e}^i)_\omega = R_{\text{ref}} \left(\int_{\check{e}_i} K \right) (\check{e}^i)_\omega = (V^\star)_\omega, \tag{100}$$

$$\left(\int_{(\check{e}_i)_\omega{}^\star} K^\star \right) (\check{e}_i)_\omega = \left(\int_{-e_i} (K^\star)_\omega \right) (\check{e}_i)_\omega = G_{\text{ref}} \left(\int_{e_i} E \right) (\check{e}_i)_\omega = (I^\star)_\omega. \tag{101}$$

These equations indicate the commutativity of the diagram shown in Figure 35. Thus, Keller–Dykhne duality corresponds to circuit duality through discretization.

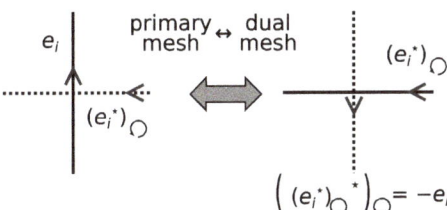

Figure 36. Interchange between a primary mesh and dual mesh.

5. Electromagnetic Duality

The electric field induced by an electric dipole has similar properties to the magnetic field created by a magnetic dipole. Such similarities can be considered as an emergent form of *electromagnetic duality*. To correctly understand electromagnetic duality, we need to clarify the difference between electric and magnetic fields. To this end, we utilize inner- and outer-oriented vectors called *polar* and *axial* vectors, respectively [78]. By using these concepts, we accurately formulate electromagnetic duality, while the role of orientation of the space in electromagnetic duality is elucidated. The analogy between electromagnetic duality and Keller–Dykhne duality is also discussed.

5.1. Preliminary

■ Polar and Axial Vectors

Here, we introduce two different kinds of vectors. Consider a line segment in a three-dimensional space. As discussed in Section 2.3, we can set an inner or outer orientation of the line. An inner-oriented line is called a *polar vector* and is represented by an arrow depicted in Figure 37a. The totality of polar vectors forms a vector space, in which we define the sum of vectors and scalar multiplication of a vector. An electric field is represented by a polar-vector field. On the other hand, an outer-oriented line can be considered as shown in Figure 37b. Such an outer-oriented line segment is called an *axial* vector. Axial vectors also form a vector space. As we saw in Section 2.3, an outer-oriented object can be represented by two-different inner-oriented ones. Here, we can represent the orientation of the space by a helix, of which the winding direction is right- or left-handed. The helix winding direction is arranged to be the same as the outer orientation of the axial vector, then the advance direction of the corresponding screw induces the inner orientation of the line as shown in Figure 38. In other words, when fingers of the left or right hand representing the helix handedness are curled in the direction of the outer orientation, the thumb points out the inner orientation. Then, an axial vector is denoted by $a = \{a_o | o \in \mathcal{O}\}$ with polar vectors satisfying $a_{-o} = -a_o$ for the set $\mathcal{O}$ of two spatial orientations. Importantly, an axial vector is a geometric object independent of the orientation of the space, although the polar vector obtained from the axial vector depends on the spatial orientation.

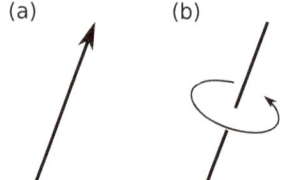

Figure 37. (a) polar vector; (b) axial vector.

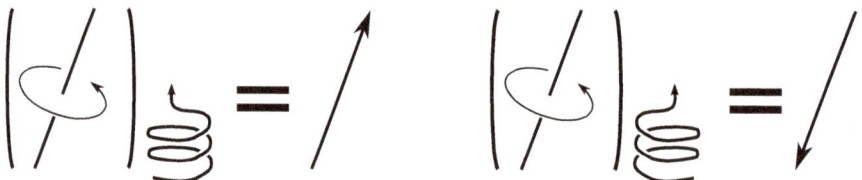

Figure 38. Representation of an axial vector by polar vectors depending on the space orientation. A spatial orientation is entered in the subscript position and its output is a polar vector which depends on the spatial orientation.

■ Magnetic Field as Axial Vectors

To explain the necessity of the axial vectors, we consider Ampère's law. The conventional image of a magnetic field induced by a current flowing along a line ($x = y = 0$) is depicted in Figure 39a, where the direction of the magnetic field is determined by the so-called right-hand rule. However, there are three unnatural points in this illustration: (i) Why is the right-hand rule required? (ii) Mirror reflection with respect to $x = 0$ or $y = 0$ does not change the current flow, but it alters the direction of the vector field. Here, mirror reflection $\mathcal{M}_x$ with respect to $x = 0$ operates as $[\mathcal{M}_x v](x,y,z) = -v^x(-x,y,z)e_x + v^y(-x,y,z)e_y + v^z(-x,y,z)e_z$ for a polar vector field $v(x,y,z) = v^x(x,y,z)e_x + v^y(x,y,z)e_y + v^z(x,y,z)e_z$. (iii) Mirror reflection with respect to $z = 0$ changes the direction of the current, but the vector field is unchanged under the operation. These three problems are resolved when we consider magnetic fields as axial vector fields. A proper illustration of the magnetic field is shown in Figure 39b, where the magnetic line is outer-oriented. In this representation, we do not need the right-hand rule. The field in Figure 39b is symmetric with respect to $x = 0$ or $y = 0$, while it is antisymmetric with respect to $z = 0$. Figure 39a is now interpreted as the right-hand component for Figure 39b.

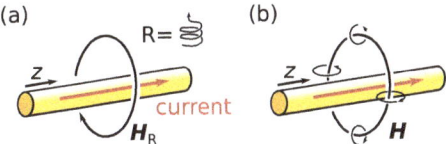

Figure 39. Ampère's law represented by (**a**) polar and (**b**) axial magnetic lines.

■ Vector Product

Another important operation that can generate an axial vector is the vector product of polar vectors. Consider the vector product between polar vectors A and B. Clearly, A and B are invariant under a mirror reflection with respect to the plane spanned by A and B. Therefore, it is natural to consider $A \times B$ as an axial vector as shown in Figure 40a. Then, the mirror symmetry is kept. The representation by polar vectors is also depicted in Figure 40b. The vector product between a polar vector and an axial vector is also defined to give a polar vector.

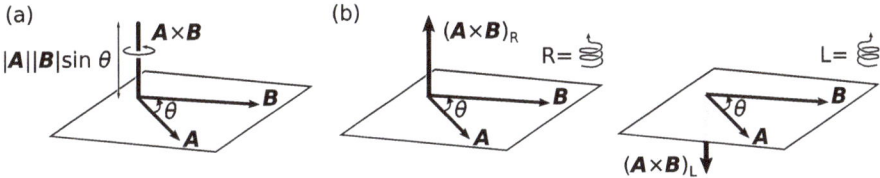

Figure 40. (**a**) axial vector obtained from a vector product of polar vectors; (**b**) its polar-vector representation.

■ Scalar and Pseudoscalar

For a scalar, we can consider an outer-oriented object. The codimension for a scalar is three in this three-dimensional space. Therefore, an outer-oriented or twisted scalar can be represented by a helix in this space. An outer-oriented scalar is often called a pseudoscalar. A pseudoscalar $\check{s}$ is represented by two scalars $\check{s} = \{\check{s}_o | o \in \mathcal{O}\}$ with $\check{s}_{-o} = -\check{s}_o$.

For example, magnetic charge density $\rho_m = \nabla \cdot B$ is a pseudoscalar because magnetic flux density B is an axial vector. Therefore, a magnetic charge should be a pseudoscalar if it exists. The difference between electric and magnetic charges is illustrated in Figure 41.

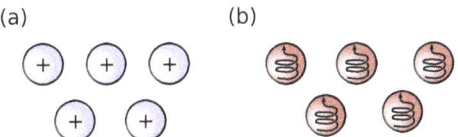

Figure 41. (**a**) electric charges as a scalar; (**b**) magnetic charges as a pseudoscalar.

5.2. Formulation of Electromagnetic Duality

■ Maxwell's Equations

Maxwell's equations can be written as

$$\nabla \times E + \frac{\partial B}{\partial t} = -J_m, \quad \nabla \cdot D = \rho_e, \qquad (102)$$

$$\nabla \times H - \frac{\partial D}{\partial t} = J_e, \quad \nabla \cdot B = \rho_m \qquad (103)$$

with electric field E, electric displacement field D, electric current density J_e, electric charge density ρ_e, magnetic field H, magnetic flux density B, magnetic current density J_m, and magnetic charge density ρ_m. While E, D, and J are polar vectors, H, B, and J_m are axial vectors. Electric and magnetic charge densities are represented by a scalar and pseudoscalar, respectively. Here, we introduce J_m and ρ_m to investigate the duality. Note that J_m and ρ_m are fictitious because a magnetic monopole does not exist. Material fields (D, H) are determined from (E, B) through the constitutive equations as described later.

■ Electromagnetic Duality Transformation

Electric and magnetic fields are represented by polar and axial vectors as shown in Figure 37, respectively. To exchange these two different types of vectors, we need to fix the orientation of the space to σ and introduce a pseudoscalar Ω^σ as $(\Omega^\sigma)_\sigma = +1$ and $(\Omega^\sigma)_{-\sigma} = -1$. The pseudoscalar Ω^σ can be represented by a helix with winding of σ as shown in Figure 42. For an axial vector a, Ω^σ converts it to the polar component as $\Omega^\sigma a = a_\sigma$.

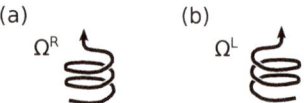

Figure 42. (a) Ω^R and (b) Ω^L.

We set a reference resistance $R_{\text{ref}}(=: 1/G_{\text{ref}})$. For a spatial orientation σ, the electromagnetic duality transformation is given by

$$E^\star = R_{\text{ref}}\Omega^\sigma H, \quad D^\star = G_{\text{ref}}\Omega^\sigma B, \tag{104}$$
$$H^\star = -G_{\text{ref}}\Omega^\sigma E, \quad B^\star = -R_{\text{ref}}\Omega^\sigma D. \tag{105}$$

Under the duality transformation, Maxwell's equations are invariant as

$$\nabla \times E^\star + \frac{\partial B^\star}{\partial t} = -J_m^\star, \quad \nabla \cdot D^\star = \rho_e^\star, \tag{106}$$
$$\nabla \times H^\star - \frac{\partial D^\star}{\partial t} = J_e^\star, \quad \nabla \cdot B^\star = \rho_m^\star \tag{107}$$

with

$$\rho_e^\star = G_{\text{ref}}\Omega^\sigma \rho_m, \quad J_e^\star = G_{\text{ref}}\Omega^\sigma J_m, \tag{108}$$
$$\rho_m^\star = -R_{\text{ref}}\Omega^\sigma \rho_e, \quad J_m^\star = -R_{\text{ref}}\Omega^\sigma J_e. \tag{109}$$

■ Duality for Constitutive Equations

Consider the relations called constitutive equations

$$D = \varepsilon E + \xi H, \tag{110}$$
$$B = \zeta E + \mu H, \tag{111}$$

where ξ and ζ are twisted. Generally, ε, ξ, ζ, and μ are tensors. Under Equations (104) and (105), the constitutive equations are transformed as

$$D^\star = \varepsilon^\star E^\star + \xi^\star H^\star, \tag{112}$$
$$B^\star = \zeta^\star E^\star + \mu^\star H^\star \tag{113}$$

with

$$\varepsilon^\star = (G_{\text{ref}})^2 \mu, \quad \xi^\star = -\zeta, \tag{114}$$

$$\zeta^\star = -\xi, \quad \mu^\star = (R_{\text{ref}})^2 \varepsilon. \tag{115}$$

■ **Self-Dual Media**

Now, we require the set of self-dual conditions $\varepsilon^\star = \varepsilon$, $\xi^\star = \xi$, $\mu^\star = \mu$, and $\zeta^\star = \zeta$. Then, the following equations should hold:

$$\varepsilon = (G_{\text{ref}})^2 \mu, \quad \xi = -\zeta. \tag{116}$$

Here, the G_{ref} satisfying Equation (116) is called the admittance Y of the medium. For scalar ε and μ, we have $Y = \sqrt{\varepsilon/\mu}$. In particular, a vacuum with $\varepsilon = \varepsilon_0$, $\mu = \mu_0$, $\xi = \zeta = 0$ is self-dual with respect to the vacuum admittance $Y_0 = \sqrt{\varepsilon_0/\mu_0}$. In circuit theory, a self-dual system did not have backscattering. This statement is also established in electromagnetic systems under certain conditions [54]. In addition, the duality transformation can be extended to continuous one and continuous self-dual symmetry leads to the helicity conservation law [55,83–88].

5.3. Analogy between Keller–Dykhne Duality and Electromagnetic Duality

Maxwell's electromagnetic theory in a four-dimensional spacetime has an analogous structure to the sheet problem discussed in Section 4. To see this analogy, we use the differential-form approach to Maxwell's equations [62,89–92]. The wedge product, exterior derivative, integral, and Hodge star are naturally extended in dimensions greater than two.

In a three-dimensional space, an electric field and magnetic field are represented by an untwisted 1-form E and twisted 1-form H, respectively. On the other hand, a magnetic flux density is denoted by an untwisted 2-form B, while an electric displacement is represented by a twisted 2-form D. Using these quantities, we define untwisted and twisted 2-forms F and G as

$$F = E \wedge dt + B, \tag{117}$$

$$G = -H \wedge dt + D, \tag{118}$$

respectively. Maxwell's equations without a source are equivalent to

$$dF = 0, \tag{119}$$

$$dG = 0, \tag{120}$$

with the four-dimensional exterior derivative d. Now, we consider a medium with a scalar permittivity ε and permeability μ, while we set $\xi = 0$ and $\zeta = 0$. The constitutive equation is given by

$$G = Y \star F \tag{121}$$

with the admittance $Y = \sqrt{\varepsilon/\mu}$ ($= 1/Z$) and the four-dimensional Hodge star operator $\star$. These equations are summarized in Figure 43. For 2-forms in the four-dimensional space, we have

$$\star \star = -\text{Id}, \tag{122}$$

with the identity operator Id. Equation (122) leads to a complex structure.

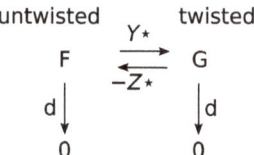

Figure 43. Structure of Maxwell's theory of electromagnetism.

Equations (119)–(121) perfectly correspond to Equations (79)–(81), respectively. When we fix an orientation ϖ of the four-dimensional spacetime, the duality transformation can be written as

$$F^\star = -R_{\text{ref}} \Omega^\varpi G, \tag{123}$$

$$G^\star = G_{\text{ref}} \Omega^\varpi F. \tag{124}$$

Then, the dual admittance is defined as

$$Y^\star = (G_{\text{ref}})^2 / Y. \tag{125}$$

The duality transformation of Equations (123) and (124) interchanges E and H in the three-dimensional space. If we set $G_{\text{ref}} = Y$, we obtain $Y^\star = Y$, which indicates the system is self-dual. As an example, a vacuum is self-dual with respect to the vacuum admittance $Y_0 = \sqrt{\varepsilon_0/\mu_0} \, (= 1/Z_0)$ with vacuum permittivity ε_0 and vacuum permeability μ_0. Electromagnetic duality can be considered as a manifestation of Poincaré duality in this spacetime [93].

6. Babinet Duality

Babinet's principle known in optics and electromagnetism relates wave-scattering problems of two complementary screens (Figure 44) [94]. Here, we call the duality appearing in Babinet's principle as *Babinet duality*. Babinet duality can be regarded as a high-frequency counterpart of Keller–Dykhne duality, which is discussed in Section 4. At first, we introduce rigorous Babinet's principle for electromagnetic waves. Then, we analyze self-dual systems in terms of Babinet duality, such as the Mushiake principle in antenna theory [13]. Finally, we discuss Babinet duality in the light of circuit duality by using a transmission-line model of metasurfaces.

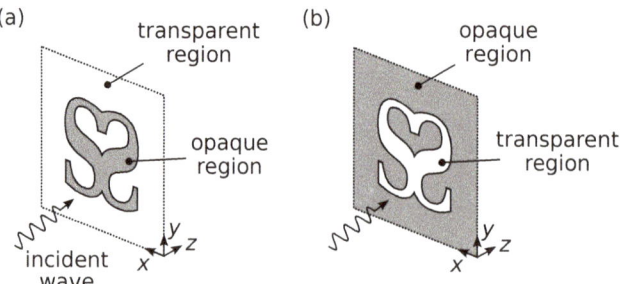

Figure 44. (**a**,**b**) two scattering problems that are dual with each other. Opaque and transparent regions are interchanged under the duality transformation. The screen in (**b**) is called the complementary screen of that in (**a**), and vice versa.

6.1. Babinet's Principle for Electromagnetic Waves

Before moving on to Babinet's principle for electromagnetic waves, we discuss the duality for electromagnetic waves radiated from planar sheets. Next, the formulated duality is utilized to derive Babinet's principle for electromagnetic waves.

■ Duality for Radiation from Planar Antennas

Here, we formulate the duality for fields radiated from nonuniform planar sheets in a vacuum, while we stress the importance of axial vectors. Consider a sheet on $z = 0$ with an electric sheet impedance $Z_e(x, y)$ and an external electric field $\tilde{E}_{ext}(x, y)$ as a voltage source. Generally, $Z_e(x, y)$ is a tensor. Following Maxwell's equations, the induced current distribution radiates electromagnetic waves. The radiated electromagnetic field is represented by $(\tilde{E}_+, \tilde{H}_+)$ and $(\tilde{E}_-, \tilde{H}_-)$ in $z \geq 0$ and $z \leq 0$, respectively. Mirror reflection with respect to $z = 0$ is expressed as $\mathcal{M}_z$ and the considered system is invariant under $\mathcal{M}_z$. First, let us see the symmetry property of electromagnetic fields on $z = 0$. The component of v perpendicular to the plane $z = 0$ is obtained by $v_n = (v \cdot e_z)e_z$. Then, the projection of v onto $z = 0$ is given by $v_t = \mathcal{P}v = v - v_n$ with $\mathcal{P} = -e_z \times e_z \times$. A polar vector p and axial vector a behave differently for $\mathcal{M}_z$ as

$$\mathcal{M}_z p = p_t - p_n, \tag{126}$$
$$\mathcal{M}_z a = -a_t + a_n. \tag{127}$$

These relations are schematically shown in Figure 45. From Equations (126) and (127), we obtain the following symmetry at any point of the plane $z = 0$:

$$\mathcal{P}\tilde{E}_+ = \mathcal{P}\tilde{E}_-, \quad e_z \cdot \tilde{E}_+ = -e_z \cdot \tilde{E}_-, \tag{128}$$
$$\mathcal{P}\tilde{H}_+ = -\mathcal{P}\tilde{H}_-, \quad e_z \cdot \tilde{H}_+ = e_z \cdot \tilde{H}_-. \tag{129}$$

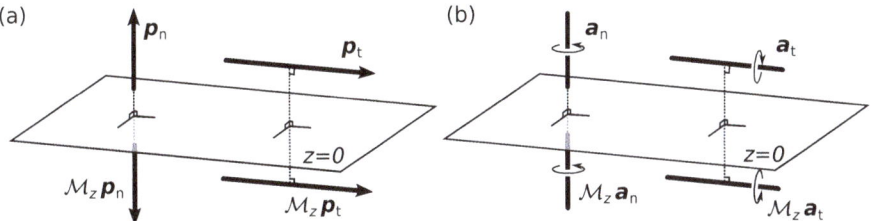

Figure 45. Mirror reflection of (**a**) polar and (**b**) axial vectors.

Next, the boundary condition on $z = 0$ is given by

$$\mathcal{P}\tilde{E}_+ = \mathcal{P}\tilde{E}_-, \tag{130}$$
$$\tilde{E}_{2D} = \mathcal{P}(\tilde{E}_+ + \tilde{E}_{ext}) = Z_e \tilde{K}_{2D}, \tag{131}$$

with the two-dimensional electric field $\tilde{E}_{2D}$ on $z = 0$, the sheet current density $\tilde{K}_{2D} = e_z \times (\tilde{H}_+ - \tilde{H}_-) = 2e_z \times \tilde{H}_+$ which is obtained from Equation (129). Equation (130) represents the continuity of the tangential electric field, while Equation (131) is equivalent to Ohm's law. The boundary conditions for $\tilde{D}$ and $\tilde{H}$ are derived from these boundary conditions [95].

Finally, we consider the duality transformation. We fix a spatial orientation σ and introduce a pseudoscalar Ω^σ satisfying $(\Omega^\sigma)_\sigma = 1$ and $(\Omega^\sigma)_{-\sigma} = -1$. The following duality transformations are considered:

1. $z \geq 0$:
$$\tilde{E}_+^\star = -Z_0 \Omega^\sigma \tilde{H}_+, \quad \tilde{H}_+^\star = Y_0 \Omega^\sigma \tilde{E}_+, \tag{132}$$

2. $z \leq 0$:
$$\tilde{E}_-^\star = Z_0 \Omega^\sigma \tilde{H}_-, \quad \tilde{H}_-^\star = -Y_0 \Omega^\sigma \tilde{E}_-, \tag{133}$$

where $Z_0 (= 1/Y_0)$ is the impedance of a vacuum. The transformed fields $(\tilde{E}_\pm^\star, \tilde{H}_\pm^\star)$ are invariant under $\mathcal{M}_z$ as shown in Figure 46. With this symmetry of Equation (129), we immediately obtain $\mathcal{P}\tilde{E}_+^\star = \mathcal{P}\tilde{E}_-^\star$ on $z = 0$. On the other hand, Equation (131) is transformed as

$$\tilde{K}_{2D}^\star = Y_e^\star \tilde{E}_{2D}^\star, \tag{134}$$

where we defined

$$\tilde{K}_{2D}^\star = 2e_z \times \tilde{H}_+^\star + \tilde{K}_{ext}^\star, \tag{135}$$
$$\tilde{K}_{ext}^\star = 2Y_0 \Omega^\sigma e_z \times \tilde{E}_{ext} \tag{136}$$

with $Z_e^\star = (Y_e^\star)^{-1}$. Here, the following general impedance inversion holds:

$$Z_e^\star J Z_e J^{-1} = \left(\frac{Z_0}{2}\right)^2 \tag{137}$$

with

$$J = \begin{bmatrix} 0 & -1 \\ 1 & 0 \end{bmatrix}.$$

Note that the impedance Z_0 can be replaced with $\sqrt{\mu/\varepsilon}$ if the screen is placed in an isotropic and homogeneous medium with permeability μ and permittivity ε. Furthermore, $\tilde{E}_{2D}^\star = \mathcal{P}\tilde{E}_+^\star$ and $\tilde{K}_{2D}^\star$ satisfy

$$\tilde{E}_{2D}^\star = \frac{Z_0}{2} \Omega^\sigma e_z \times \tilde{K}_{2D}, \tag{138}$$

$$\tilde{K}_{2D}^\star = \left(\frac{Z_0}{2}\right)^{-1} \Omega^\sigma e_z \times \tilde{E}_{2D} \tag{139}$$

on $z = 0$. Equations (137), (138), and (139) perfectly correspond to Equations (49), (43), and (44). For the dual setup with the sheet impedance $Z_e^\star$ and current source $\tilde{K}_{ext}^\star$, the radiated fields are given by $(\tilde{E}_\pm^\star, \tilde{H}_\pm^\star)$. For a scalar Z_e, the impedance inversion simplifies to

$$Z_e Z_e^\star = \left(\frac{Z_0}{2}\right)^2. \tag{140}$$

Note that Equation (140) includes the duality between the perfect electric conductor ($Z_e = 0$) and aperture ($Z_e = \infty$). In other words, the sheet-impedance model is a generalization of the binarized case, where only opaque and transparent regions were considered for screens as shown in Figure 44.

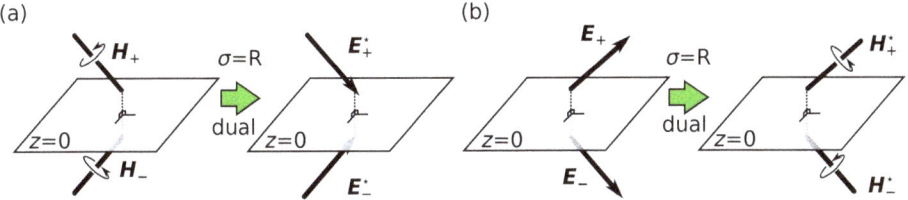

Figure 46. Duality transformation keeping the mirror symmetry. (a) $H_\pm$ to $E^\star_\pm$ and (b) $E_\pm$ to $H^\star_\pm$.

■ **Babinet's Principle for Sheet-Impedance Screens**

Here, we derive Babinet's principle by applying the previous discussion to scattering problems. We consider a scattering problem by a screen characterized with a sheet impedance of $Z_e(x,y)$ for an incident electromagnetic field $(\tilde{E}_{in}, \tilde{H}_{in})$ from $z < 0$, as shown in Figure 47a. Fields scattered by the screen are denoted by $(\tilde{E}_{s,\pm}, \tilde{H}_{s,\pm})$ for $z \geq 0$ and $z \leq 0$, respectively. These scattered fields are induced by the external electric field $\tilde{E}_{ext} = \mathcal{P}\tilde{E}_{in}(x,y,0)$ on $z = 0$.

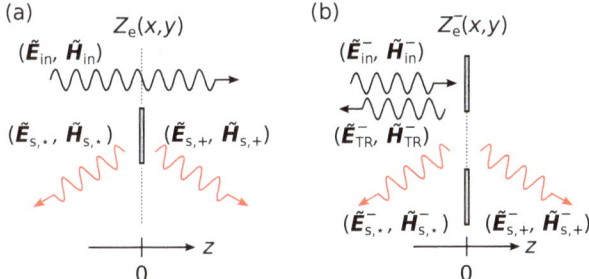

Figure 47. (a,b) two scattering problems that are dual with each other. The dual screen (b) is obtained through the impedance inversion of Equation (137).

Next, we consider the dual wave scattering by $Z_e^\star(x,y)$ for an incident wave

$$(\tilde{E}_{in}^\star, \tilde{H}_{in}^\star) = (Z_0\Omega^\sigma \tilde{H}_{in}, -Y_0\Omega^\sigma \tilde{E}_{in}), \tag{141}$$

from $z < 0$. In the dual problem, we have to inject an external current rather than apply an electric field. To this end, we virtually consider total reflection by a perfect electric conductor sheet at $z = 0$. The totally reflected field $(\tilde{E}_{TR}^\star, \tilde{H}_{TR}^\star)$ is obtained by a mirror reflection of $(\tilde{E}_{in}^\star, \tilde{H}_{in}^\star)$ in $z > 0$ with respect to $z = 0$ and a phase flip:

$$(\tilde{E}_{TR}^\star, \tilde{H}_{TR}^\star) = -(\mathcal{M}_z\tilde{E}_{in}^\star, \mathcal{M}_z\tilde{H}_{in}^\star). \tag{142}$$

Using the total reflection, we can introduce an external current as

$$\tilde{K}_{ext}^\star = -2e_z \times \tilde{H}_{in}^\star = 2Y_0\Omega^\sigma e_z \times \tilde{E}_{ext} \tag{143}$$

on $z = 0$. This virtually injected current radiates the electromagnetic field $(\tilde{E}_{s,\pm}^\star, \tilde{H}_{s,\pm}^\star)$ as shown in Figure 47b. Here, Equation (143) is identical to Equation (136). Therefore, $(\tilde{E}_{s,\pm}, \tilde{H}_{s,\pm})$ and $(\tilde{E}_{s,\pm}^\star, \tilde{H}_{s,\pm}^\star)$ are related through Equations (132) and (133), if $Z_e^\star$ satisfies Equation (137). Thus, we could relate the scattered fields in the two problems. This duality relationship is the Babinet's principle for vector waves.

Babinet's principle leads to complementary relation on transmission coefficients. Consider a normal incidence of a plane wave to a periodic screen (metasurface) with a sheet impedance $Z_e(x,y)$

on $z = 0$. The complex transmission coefficient to the transmitting mode with the same polarization is denoted by τ. In the dual situation, the incident wave of Equation (141) enters the dual metasurface $Z_e^\star(x,y)$ satisfying Equation (137). The complex electric transmission coefficient in the dual setup is represented by $\tau^\star$. Now, the following dual relation holds:

$$\tau + \tau^\star = 1. \tag{144}$$

The detailed derivation of this relation is given in Appendix B.

■ Babinet's Principle for Power Transmissions

Next, we consider the consequence of Babinet duality on power transmission spectra of periodic screens, i.e., the square of the absolute value of the complex amplitude transmission coefficients. Using the continuity equation $1 + \varrho = \tau$ with a reflection coefficient ϱ, we obtain

$$\varrho = -\tau^\star. \tag{145}$$

On the other hand, we have the energy–conservation relation

$$|\tau|^2 + |\varrho|^2 = 1, \tag{146}$$

under the following conditions: (i) periodicity of screens is smaller than the wavelength and thus energy scattering into the other diffraction modes except for the zeroth order modes is negligible, and (ii) polarization conversion is negligible ($\tau^\perp = \rho^\perp = 0$ in Appendix B). By using Equations (145) and (146), we obtain

$$|\tau|^2 + |\tau^\star|^2 = 1. \tag{147}$$

This equation indicates that the power transmission spectrum in the dual problem has the opposite shape of that in the original one.

6.2. Self-Dual Systems in Terms of Babinet Duality

In this subsection, we discuss the manifestation of self-duality in terms of Babinet duality. First, we introduce self-complementary antennas, and then we discuss the criticality of metallic checkerboard-like metasurfaces.

■ Self-Complementary Antennas

In antenna theory, it is well-known that antennas with self-complementary geometry show a frequency-independent input impedance, which is defined as the ratio of voltage and current at a feeding point of an antenna, and such antennas are called self-complementary antennas [13].

Before moving on to the self-complementary case, we introduce duality for the effective response of antennas. Consider an antenna with an electric sheet admittance $Y_e(x,y) (= Z_e(x,y)^{-1})$ on $z = 0$ as shown in Figure 48a. On a rectangular patch S, an external voltage distribution $\tilde{E}_{ext}(x,y)$ is applied. We assume that $\tilde{E}_{ext}$ is directed in the y-direction. For the dual setup, we set $Z_e^\star$ satisfying Equation (137) and an external current $\tilde{K}_{ext}^\star = 2Y_0\Omega^\sigma e_z \times \tilde{E}_{ext}$ on S in Figure 48b. In particular, when antennas are made only of perfect electric conductor ($Y_e = \infty$), the corresponding dual antennas have complementary shapes, which are obtained by interchanging metallic regions and hole regions. The input impedances of the original and dual antenna are denoted by Z_{in} and $Z_{in}^\star$, respectively. From Babinet duality, these input impedances satisfy

$$Z_{in} Z_{in}^\star = \left(\frac{Z_0}{2}\right)^2, \tag{148}$$

as derived in Appendix C. This equation means that the input impedance of the dual antenna is related to that of the original antenna.

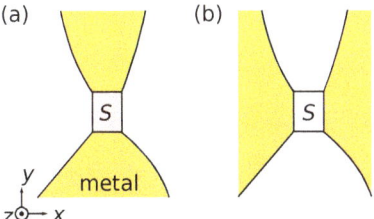

Figure 48. (**a**) antenna on $z = 0$ with voltage source on S; (**b**) dual antenna with current source on S. Note that metallic and vacant regions are interchanged through the impedance inversion.

Now, we consider self-dual antennas. An example of a self-dual antenna is realized with a self-complementary geometry as shown in Figure 49. From Equation (148) and self-dual condition $Z_{in} = Z_{in}^{\star}$, we obtain

$$Z_{in} = Z_{in}^{\star} = \frac{Z_0}{2}, \tag{149}$$

and this means that the input impedance of self-complementary antennas is independent of frequency. Thus, the principle of self-complementary antennas called the "Mushiake Principle" [96] plays an important role in designing broadband antennas [13]. Because semi-infinite free spaces in $z > 0$ and $z < 0$ are seen as two parallel transmission lines with the characteristic impedance of Z_0, Equation (149) shows that self-complementary antennas are perfectly matched to the composite impedance of $Z_0/2$.

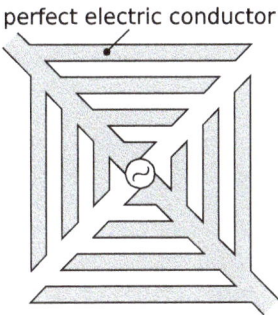

Figure 49. Example of a self-complementary antenna.

■ Critical Transition of Metallic Checkerboard-Like Metasurfaces

Here, we discuss critical behaviors of metallic checkerboard-like metasurfaces from the point of view of self-duality in Babinet duality. We assume the metasurfaces are placed in a vacuum and made of a perfect electric conductor. If we assume the periodicity of the checkerboard-like metasurfaces, we can classify them into three distinct cases as shown in Figure 50: (a) metallic patches are disconnected (disconnected phase), (b) metasurface is self-complementary, i.e., the metallic patches touch each other at ideal point contacts (self-complementary point), and (c) metallic patches are connected (connected phase). Note that structures in (c) are complementary to those in (a) if $w = w^{\star}$; therefore, the two phases are related through Babinet duality. Now, consider the transition from the disconnected phase to the connected phase. Under this transition, the checkerboard-like metasurface passes through the self-complementary point between the two phases, as shown in Figure 50. As we see below, the electromagnetic responses of checkerboard-like metasurfaces abruptly change at this self-complementary point, and this point is actually a *singular* point [44,45,47].

At first, we explain general transmission properties of checkerboard-like metasurfaces. We consider that a circularly polarized plane wave with wavevector $k = ke_z$ is normally incident on the metasurfaces, and we observe the transmission behind the metasurfaces. For the disconnected phase shown in Figure 50a, the power-transmission spectrum is typically like the lower panel of Figure 50a. The metasurfaces in the disconnected phase behave as capacitive filters, which highly transmit lower frequency components, while they resonantly reflect around the higher resonance frequency [97]. Note that the incident frequency axis is clipped at the lowest diffraction frequency c_0/a, where a is the size of the unit cell of the metasurfaces. Above the lowest diffraction frequency, the incident energy is enabled to be transmitted into higher-order diffraction modes with $k \neq \pm ke_z$. In other words, effective loss (mode-conversion loss) appears for the zeroth-order mode. To restrict our discussion to the single mode, we consider frequencies below c_0/a in the following. In addition to this constraint, thanks to the 4-fold rotational symmetry of the metasurfaces, polarization conversion is prohibited ($\tau^\perp = \rho^\perp = 0$). Then, we can use Equation (147) to obtain the power-transmission spectrum of the complementary structures: $|\tau^\star|^2 = 1 - |\tau|^2$. In other words, the connected phase shown in Figure 50c exhibits transmission spectra with the upside-down shapes to those of the disconnected phase: it highly reflects lower frequency components, while it resonantly transmits around the higher resonant frequency, as shown in the lower panel of Figure 50c [97].

Now, we derive the criticality of the self-complementary structure. At the self-complementary point shown in Figure 50b, we have

$$\tau = \tau^\star, \tag{150}$$

from the self-duality of the problem. Combining above with Equations (144) and (145), we can derive

$$\tau = \tau^\star = -\varrho = \frac{1}{2}. \tag{151}$$

Then, it is concluded that the self-complementary metasurface shows a *finite dissipation* $A = 1 - |\tau|^2 - |\varrho|^2 = 1/2$. The finite dissipation obviously contradicts the assumption that the system is made of a lossless perfect electric conductor. Note that this contradiction occurs below the diffraction frequency c_0/a. For frequencies above c_0/a, scattering into the diffraction modes is enabled, and thus A is composed of not only dissipation but also mode conversion. In addition to this contradiction, the frequency-independent behavior also contradicts Foster's reactance theorem, which states that the imaginary part of the impedance of a lossless and passive system must increase monotonically with the frequency [98]. Thus, the transmission spectra of such systems cannot be flat. Consequently, we can conclude that there is no physical solution for the ideal self-complementary point and thus the checkerboard-like metasurface at the self-complementary point is singular.

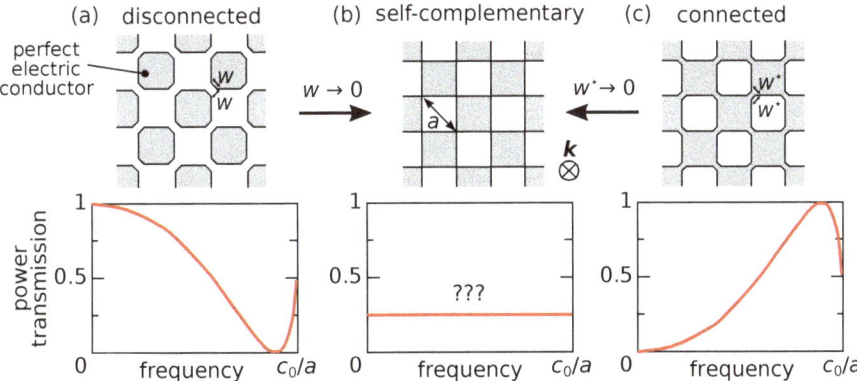

Figure 50. Metallic checkerboard-like metasurfaces and their typical power-transmission spectra. (**a**) disconnected phase; (**b**) self-complementary point; and (**c**) connected phase. The size of the unit cell is denoted by a.

Note that the *dissipative* intermediate structure in the transition between the disconnected and connected phases is physically consistent, although the *lossless* intermediate structure is singular. Such a dissipative intermediate structure, which contains dissipative elements (resistive sheets with sheet impedance of $Z_0/2$) at the connecting points of the checkerboard-like metasurfaces, can be characterized by the novel frequency-independent response with $\tau = -\varrho = 1/2$ and dissipation $A = 1/2$ [37], as shown in Figure 51. The frequency-independent response has been experimentally observed in the terahertz frequency region [38]. In addition, introducing randomness into the connectivity of the metallic patches also leads to the similar flat spectrum [42,48]. In this random case, loss A results from mode conversion due to the randomness of the metasurface structure.

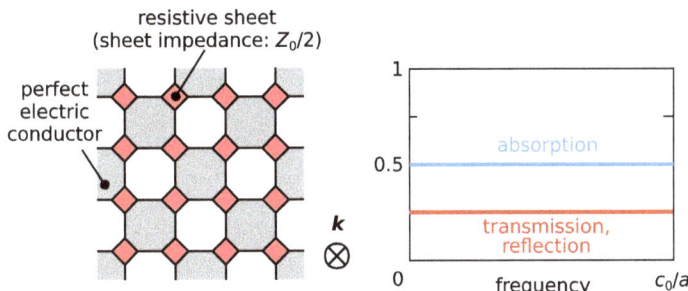

Figure 51. Resistive checkerboard-like metasurface, which is self-dual in terms of Babinet duality, and its power transmission, reflection, and absorption spectra.

The criticality of metallic checkerboard-like metasurfaces is utilized for dynamical metasurfaces to manipulate electromagnetic waves. By placing photoconductive materials like silicon or phase-change materials like vanadium dioxide (VO_2) between the metallic patches of checkerboard-like metasurfaces, researchers have realized optically tunable waveguides [99], capacitive–inductive switchable filters [50], dynamical polarizers [49], and dynamical switching of quarter-wavelength plates [52]. In addition to these experiments, dynamical planar-chirality switching is also theoretically proposed [51]. The advantage of these dynamical checkerboard-like metasurfaces is that we can achieve deep modulation of the electromagnetic characteristics of the metasurfaces because we dynamically induce phase transitions of the checkerboard-like metasurfaces.

6.3. Babinet Duality in Transmission-Line Models

Here, we consider Babinet duality in light of equivalent circuit models of metasurfaces. As discussed in Section 3, responses of metasurfaces can be described by equivalent circuit models. An electric metasurface in a vacuum can be modeled by an effective shunt impedance Z_{sh} inserted between two semi-infinite transmission lines with the characteristic impedance of Z_0 and the phase velocity of c_0 as shown in Figure 52. The complex amplitude transmission coefficient $\tau := \tilde{V}_t / \tilde{V}_{in}$ through the metasurface is written as

$$\tau = \frac{Z_{sh}}{Z_0/2 + Z_{sh}}. \tag{152}$$

On the other hand, for the dual problem with the complementary metasurface, the dual transmission coefficient τ^* is written as

$$\tau^* = \frac{Z_{sh}^*}{Z_0/2 + Z_{sh}^*} \tag{153}$$

with the effective impedance of the dual metasurface Z_{sh}^*.

Requiring the same duality relation with Equation (144):

$$\tau + \tau^* = 1, \tag{154}$$

we obtain

$$Z_{sh} Z_{sh}^* = \left(\frac{Z_0}{2}\right)^2. \tag{155}$$

This equation indicates that the effective impedance of the complementary metasurface is given by the *dual* of the effective impedance of the original metasurface with respect to the reference impedance $Z_0/2$. In other words, the response of the complementary metasurface can be described by the dual circuit to the equivalent circuit of the original metasurface.

In particular, when the shunting equivalent circuit is a self-dual circuit like the bridge circuit shown in Figure 19, i.e., $Z_{sh} = Z_{sh}^* = Z_0/2$ (see Section 2.6), we have $\tau = \tau^* = 1/2$, and the transmission and reflection characteristics are frequency independent. Thus, we can connect the frequency-independent responses of self-dual resistive checkerboard-like metasurfaces to the constant resistance property of self-dual circuits.

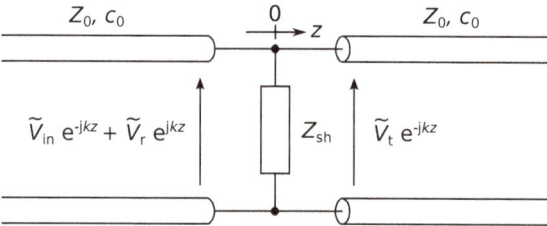

Figure 52. Transmission-line model for scattering of plane waves by an electric metasurface.

7. Conclusions

A space can be discretized with primary and dual meshes. Topological properties of the primary and dual meshes are related through Poincaré duality with each other. Especially in a $2n$-dimensional space such as a plane ($n = 1$) and spacetime ($n = 2$), the dual counterpart of an inner-oriented n-chain is given by the outer-oriented n-chain. Thus, in an even-dimensional space, duality transformation does not alter the dimension of an n-chain; rather, it changes the characteristics of orientation for the chain. This special feature of even-dimensional spaces induces the complex structure for n-chains. Inner-oriented and outer-oriented objects work as real and imaginary parts of a complex number, respectively. Due to this complex structure, the direction of an object is reversed when duality transformations are applied twice for the object. Electromagnetic duality is induced by a four-dimensional complex structure which appears from Poincaré duality. For a planar structure in the three-dimensional structure, Babinet duality holds due to the combination of electromagnetic duality and mirror symmetry. Keller–Dykhne duality is considered as an appearance of the Babinet duality at lower frequencies of direct current and alternating current. Circuit duality is interpreted as a discretized version of Keller–Dykhne duality.

Moreover, we can consider an additional invariance under duality transformation. When the system has such internal symmetry, it is called self-dual. The effective response of a self-dual system is automatically determined by the self-duality regardless of its components. This consequence of self-duality can lead to a frequency-independent input impedance and zero backscattering. Furthermore, critical response can even be predicted from self-duality, and it is leveraged to manipulate the spectra and polarization of electromagnetic waves.

In conclusion, we have unveiled the underlying geometrical structures behind various dualities in electromagnetic systems. Now, various dualities in electromagnetic systems emerge from the correspondence between quantities with two different kinds of orientations through Poincaré duality. The manifestations of self-duality in electromagnetic systems were consistently confirmed in a broad frequency range.

Author Contributions: Conceptualization, Y.N.; investigation, all authors; writing—original draft preparation, Y.N. (Sections 1, 2, 4, 5 and 7), Y.U. (Section 6), T.N. (Section 3); review and editing, all; visualization, all.; project administration, Y.N.; funding acquisition, Y.N. and T.N.

Funding: This research was funded by a grant from the Murata Science Foundation, a grant from Shimadzu Science Foundation, and JSPS KAKENHI Grant Nos. 17K17777 and 17K05075.

Acknowledgments: The authors thank Shuhei Tamate for his interpretation of the algebraic treatment of circuits, Shogo Tanimura and Masao Kitano for their fruitful discussions on foundation of electromagnetism. Shin-itiro Goto and Jacob Koenig carefully read the manuscript and gave us valuable comments.

Conflicts of Interest: The authors declare no conflict of interest.

Appendix A. General Inner-Orientation Representation for Outer Orientation

The inner-orientation representation for outer orientation is generalized for any m-dimensional space. We use the fact that an inner orientation of a vector space U can be specified by a totally antisymmetric tensor $\omega : U \times U \times \cdots \times U \to \mathbb{R}$, which is linear for all input slots and satisfies $\omega(u_1, u_2, \cdots, u_i, \cdots, u_j, \cdots, u_m) = -\omega(u_1, u_2, \cdots, u_j, \cdots, u_i, \cdots, u_m)$ for all possible $i \neq j$. For an inner-oriented vector space U, we select a positively-oriented basis $(e_1, e_2, \cdots, e_m)$ and set ω satisfying $\omega(e_1, e_2, \cdots, e_m) = 1$. If $\omega(e'_1, e'_2, \cdots, e'_m) > 0$ is satisfied, a basis $(e'_1, e'_2, \cdots, e'_m)$ has the positive orientation. Reversely, for given $\omega \neq 0$, we can specify a basis $(e_1, e_2, \cdots, e_m)$ with positive orientation as $\omega(e_1, e_2, \cdots, e_m) > 0$. Now, consider an outer-oriented linear subspace W in U with $l := \dim W < m := \dim U$. The surjective projection is denoted by $u \in U \mapsto [u] \in U/W$. Assume that an orientation of ambient space U is specified by a totally antisymmetric tensor with m slots ω_U and take a basis $([u_1], [u_2], \cdots, [u_{m-l}])$ which has positive orientation with respect to U/W. Then, we can define a totally antisymmetric tensor ω_W over W as

$$\omega_W(w_1, w_2, \cdots, w_l) := \omega_U(u_1, u_2, \cdots, u_{m-l}, w_1, w_2, \cdots, w_l), \tag{A1}$$

where $w_1, w_2, \cdots, w_l \in W$ and the first $(m-l)$-slots for the orientation tensor ω_U are contracted by an outer orientation. Thus, ω_W determines the inner orientation of W depending on the ambient-space orientation U. It is easily shown that the defined orientation of W does not depend on a specific choice of $u_1, u_2, \cdots, u_{m-l}$. The above discussion is applied to all tangent spaces smoothly, and then we can represent an outer-oriented cell as two inner-oriented cells.

Appendix B. Babinet's Principle for Transmission and Reflection Coefficients

We derive Babinet's principle for transmission and reflection coefficients for periodic screens (metasurfaces) in order to get insight into Babinet's principle under practical situations. We assume that the complex amplitude of the electric field of the incident plane wave in Figure 47a may be written as

$$\tilde{E}_{\text{in}} = \tilde{E}_{\text{in}} e_{\text{in}} \exp(-jkz), \tag{A2}$$

where e_{in} is a unit polarization vector perpendicular to e_z. Here, e_{in} can be a linear polarization like e_x, e_y or circular polarizations $e_\pm := (e_x \pm je_y)/\sqrt{2}$. Then, the scattered electric field can be expanded as

$$\tilde{E}_{s,\pm} = \left(\tilde{E}_s e_{\text{in}} + \tilde{E}_s^\perp e_{\text{in}}^\perp \right) \exp(\mp jkz) \tag{A3}$$

in the far-field region, where $e_{\text{in}}^\perp$ is the unit polarization vector orthogonal to e_{in}. While in general all diffracted waves caused by the periodic screens should also be considered, we here focus on the zeroth order modes with the wavevector $\pm k e_z$ for simplicity (for a more general formulation, see [37]). Thus, the complex amplitude transmission coefficients for the parallelly and orthogonally polarized modes τ and $\tau^\perp$ are given by

$$\tau := \frac{\tilde{E}_{\text{in}} + \tilde{E}_s}{\tilde{E}_{\text{in}}}, \tag{A4}$$

$$\tau^\perp := \frac{\tilde{E}_s^\perp}{\tilde{E}_{\text{in}}}, \tag{A5}$$

respectively. Similarly, we can write the complex amplitude reflection coefficients for the parallelly and orthogonally polarized modes ϱ and $\varrho^\perp$ as

$$\varrho := \frac{\tilde{E}_s}{\tilde{E}_{\text{in}}} = \tau - 1, \tag{A6}$$

$$\varrho^\perp := \frac{\tilde{E}_s^\perp}{\tilde{E}_{\text{in}}} = \tau^\perp, \tag{A7}$$

respectively.

Next, we consider the dual problem shown in Figure 47b. Here, we fix the spatial orientation as $\sigma = R$ for simplicity, where R represents the right-hand system. As shown in Equation (141), the complex amplitude of the electric field of the dual incident wave is given by

$$\begin{aligned}
\tilde{E}_{\text{in}}^\star &= Z_0 (\tilde{H}_{\text{in}})_R \\
&= \tilde{E}_{\text{in}} (e_z \times e_{\text{in}})_R \exp(-jkz) \\
&= \tilde{E}_{\text{in}} e_{\text{in}}^\star \exp(-jkz),
\end{aligned} \tag{A8}$$

where $e_{\text{in}}^\star := (e_z \times e_{\text{in}})_R$ is the dual incident polarization vector and $\tilde{H}_{\text{in}} = Y_0 e_z \times \tilde{E}_{\text{in}}$ is used, which can be derived from Faraday's law. According to Equation (142), the totally reflected field is written as $\tilde{E}_{\text{TR}}^\star = -\tilde{E}_{\text{in}} e_{\text{in}}^\star \exp(jkz)$. From Babinet's principle, the dual scattered electric field is given by

$$\begin{aligned}
\tilde{E}_{s,\pm}^\star &= \mp Z_0 (\tilde{H}_{s,\pm})_R \\
&= \mp Z_0 (\pm Y_0 e_z \times \tilde{E}_{s,\pm})_R \\
&= -\left[\tilde{E}_s (e_z \times e_{\text{in}})_R + \tilde{E}_s^\perp (e_z \times e_{\text{in}}^\perp)_R \right] \exp(\mp jkz) \\
&= -\left(\tilde{E}_s e_{\text{in}}^\star + \tilde{E}_s^\perp e_{\text{in}}^{\perp,\star} \right) \exp(\mp jkz),
\end{aligned} \tag{A9}$$

where $e_{in}^{\perp,\star} := (e_z \times e_{in}^\perp)_R$ is the dual orthogonal polarization vector. Finally, the dual complex transmission and reflection coefficients can be expressed as

$$\tau^\star := -\frac{\tilde{E}_s}{\tilde{E}_{in}}, \tag{A10}$$

$$\tau^{\perp,\star} := -\frac{\tilde{E}_s^\perp}{\tilde{E}_{in}}, \tag{A11}$$

$$\varrho^\star := -\frac{\tilde{E}_{in} + \tilde{E}_s}{\tilde{E}_{in}} = \tau^\star - 1, \tag{A12}$$

$$\varrho^{\perp,\star} := -\frac{\tilde{E}_s^\perp}{\tilde{E}_{in}} = \tau^{\perp,\star}. \tag{A13}$$

Note that these coefficients are defined over the dual polarization basis $e_{in}^\star$ and $e_{in}^{\perp,\star}$. If e_{in} is linearly polarized, then $e_{in}^\star$ corresponds to the orthogonal linear polarization. On the other hand, if e_{in} is circularly polarized, then $e_{in}^\star = \pm j e_{in}$ and the dual polarization state is the same as the original one up to a phase factor $\pm j$, where $+ (-)$ corresponds to the left (right) circular polarization. Here, we define handedness of circularly polarized plane waves from the receivers' side. This convention is commonly used in optics. By comparing Equations (A10) and (A11) with Equations (A4) and (A5), respectively, we finally obtain

$$\tau + \tau^\star = 1, \tag{A14}$$

$$\tau^\perp + \tau^{\perp,\star} = 0. \tag{A15}$$

Appendix C. Duality for Input Impedances of Antennas

We derive a dual relation for input impedances of antennas shown in Figure A1a,b. To simplify the discussion, we fix the spatial orientation as $\sigma = R$, where R represents the right hand. Two curves c_1 and c_2 on S are defined in Figure A1c. The total current along c_1 is calculated as

$$\tilde{I} = -2 \int_{c_2} (\tilde{H}_+)_R \cdot dr. \tag{A16}$$

The electromotive force along c_1 is given by

$$\tilde{V} = \int_{c_1} \tilde{E}_{ext} \cdot dr. \tag{A17}$$

For the dual antenna, we have a current along the c_2 direction as

$$\tilde{I}^\star = \int_{c_1} \tilde{K}_{ext}^\star \cdot (e_z \times dr)_R = 2 \int_{c_1} Y_0 \tilde{E}_{ext} \cdot dr = 2 Y_0 \tilde{V}. \tag{A18}$$

On the other hand, the electromotive force along c_2 is given by

$$\tilde{V}^\star = \int_{c_2} \tilde{E}_+^\star \cdot dr = -Z_0 \int_{c_2} (\tilde{H}_+)_R \cdot dr = \frac{Z_0}{2} \tilde{I}. \tag{A19}$$

Here, we introduce input impedances of antennas $Z_{in} = \tilde{V}/\tilde{I}$ and $Z_{in}^\star = \tilde{V}^\star/\tilde{I}^\star$, where the real part of an antenna input impedance represents electromagnetic radiation as a loss. These input impedances satisfy

$$Z_{in} Z_{in}^\star = \left(\frac{Z_0}{2}\right)^2. \tag{A20}$$

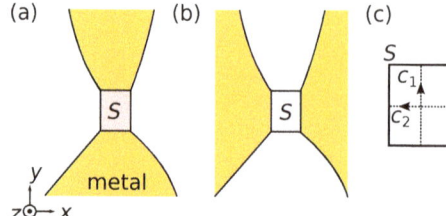

Figure A1. (a) antenna with a sheet admittance $Y_s(x,y)$ $(=:Z_s(x,y)^{-1})$ on $z=0$ is connected to voltage source $\tilde{E}_{ext}(x,y)$, which is assumed to be in the y direction, on S; (b) dual antenna with a sheet admittance $Y_s^\star(x,y)$ $(=:Z_s^\star(x,y)^{-1})$ satisfying Equation (137) on $z=0$ is connected to current source $\tilde{K}_{ext}^\star(x,y) = 2Y_0\Omega^\sigma e_z \times \tilde{E}_{ext}$ on S; (c) definition of two curves on S.

References

1. Sykes, M.F.; Essam, J.W. Exact Critical Percolation Probabilities for Site and Bond Problems in Two Dimensions. *J. Math. Phys.* **1964**, *5*, 1117–1127. [CrossRef]
2. Russo, L. On the Critical Percolation Probabilities. *Z. Wahrscheinlichkeitstheorie Verw. Geb.* **1981**, *56*, 229–237. [CrossRef]
3. Kramers, H.A.; Wannier, G.H. Statistics of the Two-Dimensional Ferromagnet. Part I. *Phys. Rev.* **1941**, *60*, 252–262. [CrossRef]
4. Vilar, H.S.D. La Dualité en électrotechnique. *L'Eclairage Electrique* **1901**, *27*, 252–259.
5. Russell, A. *A Treatise on the Theory of Alternating Currents*, 2nd ed.; Cambridge University Press: Cambridge, UK, 1914.
6. Keller, J.B. A Theorem on the Conductivity of a Composite Medium. *J. Math. Phys.* **1964**, *5*, 548–549. [CrossRef]
7. Dykhne, A.M. Conductivity of a Two-dimensional Two-phase System. *Sov. Phys. JETP* **1971**, *32*, 63–65.
8. Mendelson, K.S. A theorem on the effective conductivity of a two-dimensional heterogeneous medium. *J. Appl. Phys.* **1975**, *46*, 4740–4741. [CrossRef]
9. Heaviside, O. On the Forces, Stresses, and Fluxes of Energy in the Electromagnetic Field. *Philos. Trans. R. Soc. Lond. A* **1892**, *183*, 423–480. [CrossRef]
10. Larmor, J. A Dynamical Theory of the Electric and Luminiferous Medium. Part III. Relations with Material Media. *Philos. Trans. R. Soc. A* **1897**, *190*, 205–493. [CrossRef]
11. Rainich, G.Y. Electrodynamics in the general relativity theory. *Trans. Am. Math. Soc.* **1925**, *27*, 106–136. [CrossRef]
12. Mushiake, Y. The Input Impedances of Slit Antennas. *J. IEE Jpn.* **1949**, *69*, 87–88.
13. Mushiake, Y. *Self-Complementary Antennas: Principle of Self-Complementarity for Constant Impedance*; Springer: London, UK, 1996.
14. Babinet, M. Mémoires d'optique météorologique. *Comptes Rendus de l'Académie des Sciences* **1837**, *4*, 638–648.
15. Copson, E.T. An integral-equation method of solving plane diffraction problems. *Proc. R. Soc. A* **1946**, *186*, 100–118. [CrossRef]
16. Meixner, J. Das Babinetsche Prinzip der Optik. *Z. Naturforschg.* **1946**, *1*, 496–498. [CrossRef]
17. Leontovich, M. On a theorem in the theory of diffraction and its application to diffraction by a narrow slit of arbitrary length. *JETP* **1946**, *16*, 474–479.
18. Booker, H.G. Slot aerials and their relation to complementary wire aerials (Babinet's principle). *J. Inst. Electr. Eng. Part IIIA Radiolocat.* **1946**, *93*, 620–626. [CrossRef]
19. Kotani, M.; Takahashi, H.; Kihara, T. On the leaking of electromagnetic waves. In *Recent Developments in the Measurement of Ultrashort Waves*; Korona: Tokyo, Japan, 1948; pp. 126–134. (In Japanese)
20. Meixner, V.J. Strenge Theorie der Beugung elektromagnetischer Wellen an der vollkommen leitenden Kreisscheibe. *Z. Naturforschg.* **1948**, *3a*, 506–518. [CrossRef]

21. Aoki, T. On the Diffraction of Electromagnetic Waves by Screens and Holes of Perfect Conductors (I) On a Dual Relation between the Diffractions of Electromagnetic Waves by Screens and that by Holes of Perfect Conductors. *J. Phys. Soc. Jpn.* **1949**, *4*, 183–185. [CrossRef]
22. Aoki, T. On the Diffraction of Electromagnetic Waves by Screens and Holes of Perfect Conductors (II) Approximate Formulas and their Applications. *J. Phys. Soc. Jpn.* **1949**, *4*, 186–191. [CrossRef]
23. Bouwkamp, C.J. Diffraction Theory. *Rep. Prog. Phys.* **1954**, *17*, 35–100. [CrossRef]
24. Neugebauer, H.E.J. Extension of Babinet's Principle to Absorbing and Transparent Materials, and Approximate Theory of Backscattering by Plane, Absorbing Disks. *J. Appl. Phys.* **1957**, *28*, 302–307. [CrossRef]
25. Baum, C.E.; Singaraju, B.K. Generalization of Babinet's Principle in Terms of the Combined Field to Include Impedance Loaded Aperture Antennas and Scatterers. In *Interaction Note No.217*; Air Force Weapons Lab.: Kirtland Air Force Base, NM, USA, 1974; pp. 1–53.
26. Moore, J. Extension of the Babinet principle to scatterers with lumped impedance loads. *Electron. Lett.* **1993**, *29*, 301–302. [CrossRef]
27. Solymar, L.; Shamonina, E. *Waves in Metamaterials*; Oxford University Press: New York, NY, USA, 2009.
28. Glybovski, S.B.; Tretyakov, S.A.; Belov, P.A.; Kivshar, Y.S.; Simovski, C.R. Metasurfaces: From microwaves to visible. *Phys. Rep.* **2016**, *634*, 1–72. [CrossRef]
29. Chen, H.T.; Taylor, A.J.; Yu, N. A review of metasurfaces: physics and applications. *Rep. Prog. Phys.* **2016**, *79*, 076401. [CrossRef] [PubMed]
30. He, Q.; Sun, S.; Xiao, S.; Zhou, L. High-Efficiency Metasurfaces: Principles, Realizations, and Applications. *Adv. Opt. Mater.* **2018**, *6*, 1800415. [CrossRef]
31. Engheta, N. Circuits with Light at Nanoscales: Optical Nanocircuits Inspired by Metamaterials. *Science* **2007**, *317*, 1698–1702. [CrossRef]
32. Falcone, F.; Lopetegi, T.; Laso, M.A.G.; Baena, J.D.; Bonache, J.; Beruete, M.; Marqués, R.; Martín, F.; Sorolla, M. Babinet principle applied to the design of metasurfaces and metamaterials. *Phys. Rev. Lett.* **2004**, *93*, 197401. [CrossRef]
33. Chen, H.T.; O'Hara, J.F.; Taylor, A.J.; Averitt, R.D.; Highstrete, C.; Lee, M.; Padilla, W.J. Complementary planar terahertz metamaterials. *Opt. Express* **2007**, *15*, 1084–1095. [CrossRef]
34. Zentgraf, T.; Meyrath, T.P.; Seidel, A.; Kaiser, S.; Giessen, H.; Rockstuhl, C.; Lederer, F. Babinet's principle for optical frequency metamaterials and nanoantennas. *Phys. Rev. B* **2007**, *76*, 033407. [CrossRef]
35. Al-Naib, I.A.I.; Jansen, C.; Koch, M. Applying the Babinet principle to asymmetric resonators. *Electron. Lett.* **2008**, *44*, 1228–1229. [CrossRef]
36. Baum, C.E.; Kritikos, H.N. (Eds.) *Electromagnetic Symmetry*; Taylor & Francis: Washington, DC, USA, 1995.
37. Nakata, Y.; Urade, Y.; Nakanishi, T.; Kitano, M. Plane-wave scattering by self-complementary metasurfaces in terms of electromagnetic duality and Babinet's principle. *Phys. Rev. B* **2013**, *88*, 205138. [CrossRef]
38. Urade, Y.; Nakata, Y.; Nakanishi, T.; Kitano, M. Frequency-Independent Response of Self-Complementary Checkerboard Screens. *Phys. Rev. Lett.* **2015**, *114*, 237401. [CrossRef] [PubMed]
39. Baena, J.D.; del Risco, J.P.; Slobozhanyuk, A.P.; Glybovski, S.B.; Belov, P.A. Self-complementary metasurfaces for linear-to-circular polarization conversion. *Phys. Rev. B* **2015**, *92*, 245413. [CrossRef]
40. Urade, Y.; Nakata, Y.; Nakanishi, T.; Kitano, M. Broadband and energy-concentrating terahertz coherent perfect absorber based on a self-complementary metasurface. *Opt. Lett.* **2016**, *41*, 4472–4475. [CrossRef] [PubMed]
41. Baena, J.D.; Glybovski, S.B.; del Risco, J.P.; Slobozhanyuk, A.P.; Belov, P.A. Broadband and Thin Linear-to-Circular Polarizers Based on Self-Complementary Zigzag Metasurfaces. *IEEE Trans. Antennas Propag.* **2017**, *65*, 4124–4133. [CrossRef]
42. Takano, K.; Tanaka, Y.; Moreno, G.; Chahadih, A.; Ghaddar, A.; Han, X.L.; Vaurette, F.; Nakata, Y.; Miyamaru, F.; Nakajima, M.; et al. Energy loss of terahertz electromagnetic waves by nano-sized connections in near-self-complementary metallic checkerboard patterns. *J. Appl. Phys.* **2017**, *122*, 063101. [CrossRef]
43. Compton, R.C.; Macfarlane, J.C.; Whitbourn, L.B.; Blanco, M.M.; McPhedran, R.C. Babinet's principle applied to ideal beam-splitters for submillimetre waves. *Opt. Acta* **1984**, *31*, 515–524. [CrossRef]
44. Kempa, K. Percolation effects in the checkerboard Babinet series of metamaterial structures. *Phys. Status Solidi RRL* **2010**, *4*, 218–220. [CrossRef]

45. Edmunds, J.D.; Hibbins, A.P.; Sambles, J.R.; Youngs, I.J. Resonantly inverted microwave transmissivity threshold of metal grids. *New J. Phys.* **2010**, *12*, 063007. [CrossRef]
46. Edmunds, J.D.; Taylor, M.C.; Hibbins, A.P.; Sambles, J.R.; Youngs, I.J. Babinet's principle and the band structure of surface waves on patterned metal arrays. *J. Appl. Phys.* **2010**, *107*, 103108. [CrossRef]
47. Takano, K.; Miyamaru, F.; Akiyama, K.; Miyazaki, H.; Takeda, M.W.; Abe, Y.; Tokuda, Y.; Ito, H.; Hangyo, M. Crossover from capacitive to inductive electromagnetic responses in near self-complementary metallic checkerboard patterns. *Opt. Express* **2014**, *22*, 24787–24795. [CrossRef] [PubMed]
48. Tremain, B.; Durrant, C.J.; Carter, I.E.; Hibbins, A.P.; Sambles, J.R. The Effect of Rotational Disorder on the Microwave Transmission of Checkerboard Metal Square Arrays. *Sci. Rep.* **2015**, *5*, 16608. [CrossRef] [PubMed]
49. Nakata, Y.; Urade, Y.; Okimura, K.; Nakanishi, T.; Miyamaru, F.; Takeda, M.W.; Kitano, M. Anisotropic Babinet-Invertible Metasurfaces to Realize Transmission-Reflection Switching for Orthogonal Polarizations of Light. *Phys. Rev. Appl.* **2016**, *6*, 044022. [CrossRef]
50. Urade, Y.; Nakata, Y.; Okimura, K.; Nakanishi, T.; Miyamaru, F.; Takeda, M.W.; Kitano, M. Dynamically Babinet-invertible metasurface: A capacitive-inductive reconfigurable filter for terahertz waves using vanadium-dioxide metal-insulator transition. *Opt. Express* **2016**, *24*, 4405–4410. [CrossRef] [PubMed]
51. Urade, Y.; Nakata, Y.; Nakanishi, T.; Kitano, M. Theoretical study on dynamical planar-chirality switching in checkerboard-like metasurfaces. *EPJ Appl. Metamat.* **2017**, *4*, 2. [CrossRef]
52. Nakata, Y.; Fukawa, K.; Nakanishi, T.; Urade, Y.; Okimura, K.; Miyamaru, F. Reconfigurable Terahertz Quarter-Wave Plate for Helicity Switching Based on Babinet Inversion of an Anisotropic Checkerboard Metasurface. *Phys. Rev. Appl.* **2019**, *11*, 044008. [CrossRef]
53. Fernandez-Corbaton, I.; Fruhnert, M.; Rockstuhl, C. Objects of Maximum Electromagnetic Chirality. *Phys. Rev. X* **2016**, *6*, 031013. [CrossRef]
54. Lindell, I.V.; Sihvola, A.; Ylä-Oijala, P.; Wallén, H. Zero Backscattering From Self-Dual Objects of Finite Size. *IEEE Trans. Antennas Propag.* **2009**, *57*, 2725–2731. [CrossRef]
55. Fernandez-Corbaton, I.; Zambrana-Puyalto, X.; Tischler, N.; Vidal, X.; Juan, M.L.; Molina-Terriza, G. Electromagnetic Duality Symmetry and Helicity Conservation for the Macroscopic Maxwell's Equations. *Phys. Rev. Lett.* **2013**, *111*, 060401. [CrossRef]
56. Fernandez-Corbaton, I.; Molina-Terriza, G. Role of duality symmetry in transformation optics. *Phys. Rev. B* **2013**, *88*, 085111. [CrossRef]
57. Fernandez-Corbaton, I. Forward and backward helicity scattering coefficients for systems with discrete rotational symmetry. *Opt. Express* **2013**, *21*, 29885–29893. [CrossRef] [PubMed]
58. Paal, E.; Umbleja, M. Note on homological modeling of the electric circuits. *J. Phys. Conf. Ser.* **2014**, *532*, 012022. [CrossRef]
59. Frankel, T. *The Geometry of Physics: An Introduction*, 3rd ed.; Cambridge University Press: Cambridge, UK, 2011.
60. Bamberg, P.; Sternberg, S. *A Course in Mathematics for Students of Physics 2*; Cambridge University Press: Cambridge, UK, 1990.
61. Roman, S. *Advanced Linear Algebra*; Graduate Texts in Mathematics; Springer: New York, NY, USA, 2008.
62. Bossavit, A. Discretization of Electromagnetic Problems: The "Generalized Finite Differences" Approach. In *Handbook of Numerical Analysis*; Ciarlet, P., Ed.; Elsevier: Amsterdam, The Netherlands, 2005; Volume 13, pp. 105–197.
63. Hatcher, A. *Algebraic Topology*; Cambridge University Press: Cambridge, UK, 2001.
64. Lin, P. A Topological Method of Generating Constant Resistance Networks. *IEEE Trans. Circuit Theory* **1967**, *14*, 172–179. [CrossRef]
65. Feynman, R.; Leighton, R.; Sands, M. *The Feynman Lectures on Physics, Vol. II*; Addison-Wesley Publishing Company: Boston, MA, USA, 1965.
66. Lindell, I.V.; Sihvola, A.H. Duality Transformation for Nonreciprocal and Nonsymmetric Transmission Lines. *IEEE Trans. Microw. Theory Tech.* **1997**, *45*, 129–131. [CrossRef]
67. Heaviside, O. Electromagnetic induction and its propagation. *Electrician* **1887**, *19*, 79–81.
68. Yu, N.; Capasso, F. Flat optics with designer metasurfaces. *Nat. Mater.* **2014**, *13*, 139–150. [CrossRef]
69. Yu, N.; Genevet, P.; Kats, M.A.; Aieta, F.; Tetienne, J.P.; Capasso, F.; Gaburro, Z. Light Propagation with Phase Discontinuities: Generalized Laws of Reflection and Refraction. *Science* **2011**, *334*, 333–337. [CrossRef]

70. Kang, M.; Feng, T.; Wang, H.T.; Li, J. Wave front engineering from an array of thin aperture antennas. *Opt. Express* **2012**, *20*, 15882–15890. [CrossRef]
71. Huang, L.; Chen, X.; Mühlenbernd, H.; Li, G.; Bai, B.; Tan, Q.; Jin, G.; Zentgraf, T.; Zhang, S. Dispersionless phase discontinuities for controlling light propagation. *Nano Lett.* **2012**, *12*, 5750–5755. [CrossRef]
72. Lin, D.; Fan, P.; Hasman, E.; Brongersma, M.L. Dielectric gradient metasurface optical elements. *Science* **2014**, *345*, 298–302. [CrossRef]
73. Pfeiffer, C.; Grbic, A. Metamaterial Huygens' Surfaces: Tailoring Wave Fronts with Reflectionless Sheets. *Phys. Rev. Lett.* **2013**, *110*, 197401. [CrossRef] [PubMed]
74. Kuester, E.F.; Mohamed, M.A.; Piket-May, M.; Holloway, C.L. Averaged transition conditions for electromagnetic fields at a metafilm. *IEEE Trans. Antennas Propag.* **2003**, *51*, 2641–2651. [CrossRef]
75. Selvanayagam, M.; Eleftheriades, G.V. Circuit Modeling of Huygens Surfaces. *IEEE Antennas Wirel. Propag. Lett.* **2013**, *12*, 1642–1645. [CrossRef]
76. Needham, T. *Visual Complex Analysis*; Oxford University Press: Oxford, UK, 1997.
77. Perrins, W.T.; McPhedran, R.C. Metamaterials and the homogenization of composite materials. *Metamaterials* **2010**, *4*, 24–31. [CrossRef]
78. Burke, W.L. *Applied Differential Geometry*; Cambridge University Press: Cambridge, UK, 1985.
79. Bossavit, A. Whitney forms: A class of finite elements for three-dimensional computations in electromagnetism. *IEE Proc. A* **1988**, *135*, 493–50. [CrossRef]
80. Tarhasaari, T.; Kettunen, L.; Bossavit, A. Some realizations of a discrete Hodge operator: A reinterpretation of finite element techniques. *IEEE Trans. Magn.* **1999**, *35*, 1494–1497. [CrossRef]
81. Bossavit, A. A uniform rationale for Whitney forms on various supporting shapes. *Math. Comput. Simul.* **2010**, *80*, 1567–1577. [CrossRef]
82. Tsukerman, I. Effective parameters of metamaterials: A rigorous homogenization theory via Whitney interpolation. *J. Opt. Soc. Am. B* **2011**, *28*, 577–586. [CrossRef]
83. Calkin, M.G. An Invariance Property of the Free Electromagnetic Field. *Am. J. Phys.* **1965**, *33*, 958–960. [CrossRef]
84. Deser, S.; Teitelboim, C. Duality transformations of Abelian and non-Abelian gauge fields. *Phys. Rev. D* **1976**, *13*, 1592–1597. [CrossRef]
85. Drummond, P.D. Dual symmetric Lagrangians and conservation laws. *Phys. Rev. A* **1999**, *60*, R3331–R3334. [CrossRef]
86. Cameron, R.P.; Barnett, S.M. Electric-magnetic symmetry and Noether's theorem. *New J. Phys.* **2012**, *14*, 123019. [CrossRef]
87. Barnett, S.M.; Cameron, R.P.; Yao, A.M. Duplex symmetry and its relation to the conservation of optical helicity. *Phys. Rev. A* **2012**, *86*, 013845. [CrossRef]
88. Bliokh, K.Y.; Bekshaev, A.Y.; Nori, F. Dual electromagnetism: Helicity, spin, momentum and angular momentum. *New J. Phys.* **2013**, *15*, 033026. [CrossRef]
89. Deschamps, G. Electromagnetics and Differential Forms. *Proc. IEEE* **1981**, *69*, 676–696. [CrossRef]
90. Hehl, F.W.; Obukhov, Y.N. *Foundations of Classical Electrodynamics: Charge, Flux, and Metric*; Birkhäuser: Boston, MA, USA, 2003.
91. Kitano, M. Reformulation of Electromagnetism with Differential Forms. In *Trends in Electromagnetism—From Fundamentals to Applications*; Barsan, V., Lungu, R.P., Eds.; IntechOpen: London, UK, 2012; Chapter 2, pp. 21–44.
92. Gratus, J. A pictorial introduction to differential geometry, leading to Maxwell's equations as three pictures. *arXiv* **2017**, arXiv:1709.08492.
93. Teixeira, F.L. Lattice Maxwell's Equations. *Prog. Electromagn. Res.* **2014**, *148*, 113–128. [CrossRef]
94. Jackson, J.D. *Classical Electrodynamics*, 3rd ed.; Wiley: New York, NY, USA, 1998.
95. Yeh, C. Boundary conditions in electromagnetics. *Phys. Rev. E* **1993**, *48*, 1426–1427. [CrossRef]
96. Singh, R.; Rockstuhl, C.; Menzel, C.; Meyrath, T.P.; He, M.; Giessen, H.; Lederer, F.; Zhang, W. Spiral-type terahertz antennas and the manifestation of the Mushiake principle. *Opt. Express* **2009**, *17*, 9971–9980. [CrossRef]
97. Ulrich, R. Far-infrared properties of metallic mesh and its complementary structure. *Infrared Phys.* **1967**, *7*, 37–55. [CrossRef]

98. Foster, R.M. A reactance theorem. *Bell Syst. Tech. J.* **1924**, *3*, 259–267. [CrossRef]
99. González-Ovejero, D.; Martini, E.; Loiseaux, B.; Tripon-Canseliet, C.; Mencagli, M.; Chazelas, J.; Maci, S. Basic Properties of Checkerboard Metasurfaces. *IEEE Antennas Wirel. Propag. Lett.* **2014**, *14*, 406–409. [CrossRef]

© 2019 by the authors. Licensee MDPI, Basel, Switzerland. This article is an open access article distributed under the terms and conditions of the Creative Commons Attribution (CC BY) license (http://creativecommons.org/licenses/by/4.0/).

Perspective

Optical Helicity and Optical Chirality in Free Space and in the Presence of Matter

Lisa V. Poulikakos [1], Jennifer A. Dionne [1] and Aitzol García-Etxarri [2,3,*]

[1] Department of Materials Science and Engineering, Stanford University, 496 Lomita Mall, Stanford, CA 94305, USA
[2] Donostia International Physics Center and Centro de Fisica de Materiales CSIC-UPV/EHU, Paseo Manuel de Lardizabal 4, 20018 Donostia-San Sebastian, Spain
[3] IKERBASQUE, Basque Foundation for Science, Maria Diaz de Haro 3, 48013 Bilbao, Spain
* Correspondence: aitzolgarcia@dipc.org

Received: 17 July 2019; Accepted: 15 August 2019; Published: 3 September 2019

Abstract: The inherently weak nature of chiral light–matter interactions can be enhanced by orders of magnitude utilizing artificially-engineered nanophotonic structures. These structures enable high spatial concentration of electromagnetic fields with controlled helicity and chirality. However, the effective design and optimization of nanostructures requires defining physical observables which quantify the degree of electromagnetic helicity and chirality. In this perspective, we discuss optical helicity, optical chirality, and their related conservation laws, describing situations in which each provides the most meaningful physical information in free space and in the context of chiral light–matter interactions. First, an instructive comparison is drawn to the concepts of momentum, force, and energy in classical mechanics. In free space, optical helicity closely parallels momentum, whereas optical chirality parallels force. In the presence of macroscopic matter, the optical helicity finds its optimal physical application in the case of lossless, dual-symmetric media, while, in contrast, the optical chirality provides physically observable information in the presence of lossy, dispersive media. Finally, based on numerical simulations of a gold and silicon nanosphere, we discuss how metallic and dielectric nanostructures can generate chiral electromagnetic fields upon interaction with chiral light, offering guidelines for the rational design of nanostructure-enhanced electromagnetic chirality.

Keywords: optical chirality; optical helicity; nanophotonics; plasmonics; parity symmetry; time symmetry

1. Introduction

Chiral electromagnetic fields, exhibiting left- or right-handedness, have the ability to interact selectively with matter. In particular, the chiral molecular building blocks of biological matter, e.g., proteins and amino acids, exhibit distinct interactions with left- or right-handed light, thus enabling the visualization of the role of chirality in natural processes [1]. Additionally, the selectivity and sensitivity of chiral light–matter interactions has shown promise in a variety of technological applications. The development of pharmaceuticals faces the challenge of heterochirality, where an unbiased chemical reaction forms equal quantities of left- and right-handed products. Importantly, their biochemical interaction with patients can range from the desired pharmaceutical treatment to harmful side-effects by simply interchanging molecular chirality [2]. Thus, chiral light has been proposed as a non-invasive method to bias chemical reactions toward products of a single handedness [3–5] or to detect and separate molecules based on their chirality [6,7]. The chirality of light has additionally found potential in optical information storage and transfer [8,9], with the ability to increase capacity and selectivity.

The interaction between chiral electromagnetic plane waves, such as circularly polarized light (CPL), and matter is inherently limited in sensitivity due to their bounded spatial distribution. Rapidly-evolving research efforts in the field of chiral nanophotonics aim to address this challenge by the tailored design of metallic [10–28] and dielectric [29–38] nanostructures, arranged periodically in sub-wavelength metamaterials and metasurfaces, or in colloidal dispersions, achieving highly concentrated electromagnetic chirality in their evanescent field (see also review articles [25,39–42]). However, the rational design of enhanced electromagnetic chirality in the presence of matter requires the definition of physical observables by which to quantify the chirality of light. For this, the *optical helicity* and the *optical chirality* have been proposed, where each quantity has been altered from its free-space form to account for interactions with matter. While closely related, the optical helicity and the optical chirality differ in their physical meaning, and their application in the presence of matter is subtle yet distinct. This perspective performs a comprehensive comparison of each quantity and its physical significance in free space and in the presence of matter.

First, the optical helicity and optical chirality are introduced in the context of rotating vector fields and chiral symmetries, while the physical significance of each quantity in free space is discussed in analogy to the relationship between momentum and force in classical mechanics. Subsequently, the implications of optical helicity, optical chirality and their respective conservation laws upon interaction with microscopic and macroscopic matter are considered with a particular focus on the role of matter-induced losses. For this, an additional, physically-relevant comparison to energy and momentum conservation in classical mechanics is provided and the physical observables arising from the conservation law of optical chirality in lossy, dispersive media are discussed. Finally, we apply these observables to elucidate the physical mechanisms of chiral light–matter interactions in artificial nanostructures, where the distinct cases of metallic and dielectric nanoparticles are analyzed numerically. In particular, the chiral electromagnetic fields generated by gold and silicon nanospheres with 75 nm radius are considered, demonstrating in both cases that achiral, linearly polarized excitation does not yield a net electromagnetic chirality, while chiral excitation with left- and right-handed CPL results in mirror-symmetric optical chirality flux spectra.

2. Rotating and Handed Vector Fields

A rotating vector field is handed when its motion exhibits a non-zero component parallel to the rotational axis. This vector field property, found in a variety of natural phenomena [43,44], can be quantified by the *helicity*, a pseudoscalar resulting from projection of the angular momentum vector onto the linear momentum vector [45,46]. In fluid dynamics, the helicity is obtained from the projection of the fluid velocity $\mathbf{v}$ onto its curl, also known as vorticity, $\nabla \times \mathbf{v}$ [47]:

$$\mathcal{H}_{\text{fluid}} = \int \mathbf{v} \cdot (\nabla \times \mathbf{v}) d^3x. \tag{1}$$

In plasma physics, the magnetic helicity:

$$\mathcal{H}_{\text{magnetic}} = \int \mathbf{A} \cdot (\nabla \times \mathbf{A}) d^3x, \tag{2}$$

can be employed for the topological classification of a magnetic induction field $\mathbf{B} = \nabla \times \mathbf{A}$, with vector potential $\mathbf{A}$ [48], where integration over all of space results in gauge invariance of Equation (2) [43,49]. In contrast to Equations (1) and (2), single-particle helicity has been quantified in quantized systems where, for instance, photon helicity amounts to ±1 [43,46,50].

In classical electrodynamics, where Maxwell's equations describe the relationship between electric and magnetic fields, Equation (2) can be extended to define the *optical helicity* [43]:

$$\mathcal{H}_{\text{optical}} = \frac{1}{2} \int [\sqrt{\frac{\epsilon_0}{\mu_0}} \mathbf{A} \cdot (\nabla \times \mathbf{A}) + \sqrt{\frac{\mu_0}{\epsilon_0}} \mathbf{C} \cdot (\nabla \times \mathbf{C})] d^3x, \tag{3}$$

where **C** is the electric pseudovector potential with $\mathbf{E} = -\nabla \times \mathbf{C}$ for electric field **E** in free space [43,49]. While the vector and pseudovector potentials **A** and **C** are gauge variant, upon integration over all of space in Equation (3), only their gauge-invariant transverse components are non-zero [43]. The integrand of Equation (3), termed *optical helicity density*, is the lowest-order term in an infinite set of conserved quantities [49,51], where higher orders are obtained by mapping the magnetic and electric vector potentials onto their curls: $\mathbf{A} \to \nabla \times \mathbf{A}$ and $\mathbf{C} \to \nabla \times \mathbf{C}$ [52,53]. The first-order transformation in this series yields the *optical chirality density*, a quantity identified as physically significant in the study of chiral light–matter interactions [54]. The corresponding volume-integrated *optical chirality* is written as [54,55]:

$$\Xi = \int \left[\frac{\epsilon_0}{2}\mathbf{E} \cdot (\nabla \times \mathbf{E}) + \frac{1}{2\mu_0}\mathbf{B} \cdot (\nabla \times \mathbf{B})\right] d^3x. \tag{4}$$

While both the optical helicity (Equation (3)) and optical chirality (Equation (4)) provide information on the handedness of electromagnetic fields, they are physically distinct quantities, exhibiting a proportionality in the case of monochromatic electromagnetic fields in free space [43].

3. Physical Significance of Optical Helicity and Optical Chirality in Free Space

Noether's theorem [56] states that a conserved quantity arises in the dynamic equations of any continuous symmetry of a nondissipative system. As demonstrated by Calkin in 1965 [57], the optical helicity density h (integrand of Equation (3)) is the conserved quantity related to *electromagnetic duality symmetry* in free space, where electromagnetic duality describes a transformation between electric and magnetic fields written as: $\mathbf{E} \to \mathbf{E}_\theta = \mathbf{E}\cos\theta - \mathbf{H}\sin\theta$ and $\mathbf{H} \to \mathbf{H}_\theta = \mathbf{E}\sin\theta + \mathbf{H}\cos\theta$ [50,57].

Concurrently in 1964, Lipkin utilized Maxwell's equations to identify a new conservation law, naming the conserved quantity *Lipkin's zilch* [55]. It was not until 2010 when the physical significance of Lipkin's zilch was identified by Tang and Cohen as the local density of electromagnetic chirality, now known as the optical chirality density χ (integrand of Equation (4)).

The optical helicity density h, the optical chirality density χ and their respective flux densities **Φ** and **F**, follow formally analogous continuity equations in free space, as indicated in Table 1. This congruence along with the ability of both the optical helicity and optical chirality to describe the handedness of electromagnetic fields opens the question on how to distinguish these quantities, as discussed below.

Table 1. Conservation laws of optical helicity (**left column**) and optical chirality (**right column**) in free space with optical helicity density h, optical helicity flux density **Φ**, optical chirality density χ, optical chirality flux density **F**, electric vector potential **C**, magnetic vector potential **A**, electric field **E**, and magnetic induction field **B**. ϵ_0 and μ_0 represent the free-space electric permittivity and magnetic permeability.

Optical Helicity Conservation in Free Space	Optical Chirality Conservation in Free Space
$h = \frac{1}{2}\left[\sqrt{\frac{\epsilon_0}{\mu_0}}\mathbf{A} \cdot (\nabla \times \mathbf{A}) + \sqrt{\frac{\mu_0}{\epsilon_0}}\mathbf{C} \cdot (\nabla \times \mathbf{C})\right]$	$\chi = \frac{\epsilon_0}{2}\mathbf{E} \cdot (\nabla \times \mathbf{E}) + \frac{1}{2\mu_0}\mathbf{B} \cdot (\nabla \times \mathbf{B})$
$\mathbf{\Phi} = \frac{1}{2}\left[\sqrt{\frac{\epsilon_0}{\mu_0}}\mathbf{A} \times (\nabla \times \mathbf{C}) - \frac{1}{c}\mathbf{C} \times (\nabla \times \mathbf{A})\right]$	$\mathbf{F} = \frac{1}{2}\left[\mathbf{E} \times (\nabla \times \mathbf{B}) - \mathbf{B} \times (\nabla \times \mathbf{E})\right]$
$\frac{\delta h}{\delta t} + \frac{1}{\mu_0}\nabla \cdot \mathbf{\Phi} = 0$	$\frac{\delta \chi}{\delta t} + \frac{1}{\mu_0}\nabla \cdot \mathbf{F} = 0$

The physical significance of the optical helicity density has been described with the help of dimensional analysis, as it has units of angular momentum density $\left(\frac{\text{N m s}}{\text{m}^3}\right)$ [43,52,58]. Here, we extend this dimensional analysis to the optical chirality density χ with units of force density $\left(\frac{\text{N}}{\text{m}^3}\right)$. Their respective units of angular momentum and force invite a qualitative comparison to the relationship between momentum and force in classical mechanics, illustrated in Table 2.

Table 2. **Top**: The relationship between force and momentum for linear and rotational motion in classical mechanics, for linear momentum **p**, force **F**, angular momentum **L**, and torque τ. **Bottom**: The relationship between the optical helicity density h and the optical chirality density χ in classical electrodynamics.

Physical Significance	Fundamental	→	Observable
	Classical Mechanics		
Linear Motion: $\frac{d\mathbf{p}}{dt} = \mathbf{F}$	Linear Momentum [N s]	$\xrightarrow{\frac{d}{dt}}$	Force [N]
Rotational Motion: $\frac{d\mathbf{L}}{dt} = \tau$	Angular Momentum [N m s]	$\xrightarrow{\frac{d}{dt}}$	Torque [N m]
	Classical Electrodynamics		
Handed Motion: $h \xrightarrow{\nabla \times} \chi$	Optical Helicity Density [$\frac{\text{N m s}}{\text{m}^3}$]	$\xrightarrow{\nabla \times}$	Optical Chirality Density [$\frac{\text{N}}{\text{m}^3}$]

For a closed system with time-invariant mass, Newton's second law states that the net force exerted on the system is equal to the time derivative of the linear momentum (first row in Table 2). Similarly, for rotational motion in classical mechanics, the angular momentum is obtained from the time derivative of the torque (second row in Table 2). Equivalent information can, therefore, be obtained from the conservation of momentum and the conservation of force. However, momentum conservation, from which force conservation can be derived, is more generally valid. In contrast, for practical applications, forces lend themselves more easily to measurement and observation [59,60].

Inspired by classical mechanics, we draw a similar comparison for optical helicity and optical chirality (third row in Table 2) [61,62]. While the vector potentials inherent to the optical helicity are not directly physically observable, the optical chirality depends only on uniquely defined and observable field quantities derived from Maxwell's equations. In contrast, the optical chirality is a higher order transformation of the optical helicity. Thus, while both the optical helicity density h and the optical chirality density χ can provide information on the local handedness of an electromagnetic field in free space, h is the more fundamental quantity, while χ is more suitable for experimental observation.

Chiral Symmetries in Electromagnetism

After establishing the optical chirality density and flux as physically observable quantities which describe electromagnetic chirality, this section applies symmetry relations to illustrate how these quantities represent the chirality of electromagnetic fields [61]. A chiral system exhibits *parity odd* and *time even* symmetries [54,63]. For a function $f(x, y, z)$ with spatial coordinates x, y, and z, parity transformation occurs by inversion of the spatial coordinates through the origin. Specifically, f is parity odd when $f(x, y, z) = -f(-x, -y, -z)$ [63,64]. In addition, a function f is time even when inversion of the temporal coordinate t results in $f(t) = f(-t)$ [63,64]. Thus, chiral systems can take on two left- and right-handed mirror-symmetric forms. For chiral electromagnetic quantities χ and **F** (as defined in Table 1), the parity and time symmetries are noted in Table 3, resulting from the parity odd and time even symmetry of the electric (**E**) field and the parity even and time odd symmetry of the magnetic induction (**B**) field. To elucidate their physical origin, we now construct the symmetry relations of the **E** and **B** fields from source charges and currents.

Table 3. Parity and time symmetries of the optical chirality density χ and the optical chirality flux density **F**.

Physical Quantity	Tensor Rank	Parity Symmetry	Time Symmetry
Optical Chirality Density χ	3	Odd (pseudoscalar)	Even
Optical Chirality Flux Density **F**	1	Even (pseudovector)	Even

A static electric field is induced by interaction between positive and negative point charges (Figure 1a). For spatial coordinates (x, y, z), an electric field vector $\mathbf{E} = (E, 0, 0)$ is induced by a positive point charge at $(-x, 0, 0)$ and a negative point charge at $(+x, 0, 0)$. Parity inversion directs the electric field vector along $-x$, whereupon the parity-odd electric field vector becomes $\mathbf{E}_{\text{parity inversion}} = (-E, 0, 0)$. The time-even symmetry of $\mathbf{E}$ results from invariance of the polarity of the point charges upon time reversal.

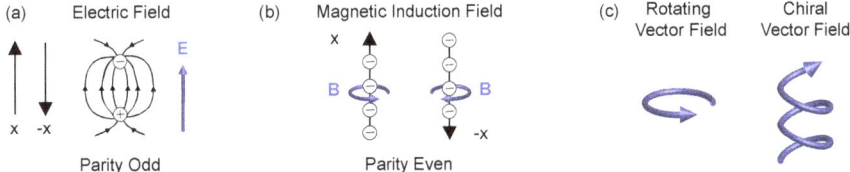

Figure 1. Illustration of parity symmetry for an electric field (**E**) arising between positive and negative point charges (**a**) and the magnetic induction field (**B**) arising from a steady state current (**b**). (**c**) Illustration of the distinction between a rotating vector field and a chiral vector field, where rotational motion has a component along the axis of rotation.

Figure 1b illustrates how a static magnetic induction field **B** is induced by a temporally steady-state current. While a current along $+x$ results in a clockwise rotation of **B**, parity inversion, directing the current along $-x$, leads to anti-clockwise rotation of **B**. Thus, **B** is parity even due to the parity-odd symmetry of the curl operator [64]. Time inversion transforms the rotation of **B** from clockwise to anti-clockwise as the current flows backwards temporally. With time invariance of the curl operator, the **B** field is, therefore, time odd. Building on these fundamental examples, the parity and time symmetries of chiral electromagnetic quantities, as noted in Table 1, are now discussed.

Regarding *parity symmetry*, the optical chirality density χ (Table 1) is a parity-odd pseudoscalar, attributed to the scalar product of a vector (the **E** field or the curl of the **B** field) and a pseudovector (the **B** field or the curl of the **E** field). In contrast, the optical chirality flux density **F** (Table 1) is a parity-even pseudovector, its flux integral thus being parity odd. As shown in Figure 1c, rotating vector fields, e.g., the parity-even fields ($\nabla \times \mathbf{E}$) or **B**, do not exhibit the symmetries of a chiral quantity [65], as their signs remain invariant upon parity inversion. Thus, a parity-odd pseudoscalar is obtained from the product of these rotating fields with their related vector fields **E** or ($\nabla \times \mathbf{B}$), respectively [61].

We now describe the *time symmetry* of a chiral electromagnetic system by comparison to a right-handed helical chiral object, the handedness of which is invariant upon forwards or backwards motion in time. Similarly, a chiral electromagnetic field exhibits rotational motion about its axis and the handedness of this motion remains unchanged under time reversal. This time-even property of chiral electromagnetic fields is seen in the optical chirality density χ due to the projection of the **E** or **B** fields onto their respective curl [61].

4. Physical Significance of Optical Helicity and Chirality upon Interaction with Matter

The free-space definitions of optical helicity and optical chirality (Table 1), require further consideration in systems where light interacts with matter. In general, the presence of matter breaks duality symmetry and the conservation of optical helicity, as predicted by Noether's theorem [56], is no longer valid. In contrast, the conservation law of optical chirality in the presence of matter requires no such restrictions in the electric and magnetic fields or the material properties. Indeed, matter has been identified as a source or sink of optical chirality [54,66,67]. While both conservation laws of optical helicity and optical chirality have been modified from their free-space form to account for interactions with matter [44,50,68–71], this section discusses the distinction in their physical relevance and applicability.

To better differentiate optical helicity and optical chirality conservation in the presence of matter, we again draw a qualitative comparison between classical mechanics and classical electrodynamics. Figure 2a illustrates the elastic (left panel) and inelastic (right panel) collision of two moving objects with masses m_1, m_2 and velocities $\mathbf{v}_1$, $\mathbf{v}_2$, respectively. For elastic collision (left), both the linear momentum $\mathbf{p} = m\mathbf{v}$ and the mechanical energy, specifically the kinetic energy, of the system are conserved at times t_1 before the collision and t_2 after the collision. In contrast, while the inelastic collision (right) conserves the linear momentum $\mathbf{p}$ of the system, the mechanical energy is no longer conserved as a portion is converted into other forms of energy (predominantly heat) [59]. Importantly, the comparison of elastic and inelastic collisions in Figure 2a reveals how the dissipation of kinetic energy in the inelastic collision is only represented by energy conservation and is not accounted for by momentum conservation. Momentum and energy conservation can thus provide equivalent information in the lossless case of an elastic collision. However, their utility differs in the dissipative case of an inelastic collision.

While physically distinct, the example of elastic and inelastic collisions in classical mechanics is instructive for the qualitative understanding of optical helicity and optical chirality conservation in matter. The presence of microscopic sources, in the form of point charges and currents, breaks duality symmetry due to the existence of electric charges and the absence of magnetic charges in matter [64]. Under constraint of the divergence-free transverse component of the current density, the conservation law of optical helicity has been reformulated to account for the presence of microscopic sources [44,70]. However, as this definition depends on vector potentials, where non-locality is circumvented by restricting the fields to their transverse components, the resulting source term is not directly physically observable. In contrast, the conservation law of optical chirality in the presence of microscopic material sources has been defined as a direct consequence of Maxwell's equations, resulting in a source term composed solely of physical, observable quantities [54]:

$$\frac{\delta \chi}{\delta t} + \frac{1}{\mu_0} \nabla \cdot \mathbf{F} = -\frac{1}{2}(\mathbf{j}_0 \cdot \nabla \times \mathbf{E} + \mathbf{E} \cdot \nabla \times \mathbf{j}_0), \tag{5}$$

where the source term arising from microscopic matter is shown on the right-hand side of Equation (5) and $\mathbf{j}_0$ is the primary current density.

Figure 2b illustrates the physical relevance of optical helicity and optical chirality conservation in the presence of macroscopic matter, in systems free of primary sources. The left panel of Figure 2b shows chiral light interacting with a piecewise homogeneous, isotropic medium with constant ϵ_i/μ_i over all material domains i [50]. Under these conditions, duality symmetry holds and both optical helicity and optical chirality are conserved [50]. Duality symmetry can also be induced in cylindrically symmetric dielectric objects and collections of objects under specific excitation conditions [72–74], also leading to helicity conservation. The conservation of optical helicity has proven instrumental in the analysis of the conversion of CPL into light with orbital angular momentum (OAM) [75–78]; it has also enabled generation of enhanced optical mirages from dual nanospheres [79,80] and an improved, elegant understanding of the interaction of vortex beams with well-defined helicity and macroscopic matter [81].

The right panel of Figure 2b represents a lossy, dispersive medium interacting with chiral light. In this case, the presence of matter breaks duality symmetry in the studied system and the optical helicity is no longer conserved in its physically observable form. In contrast, the conservation law of optical chirality can be extended from its free-space form (Table 1) to account for the presence of lossy, dispersive media, as outlined below in Section 4.1 [66,67]. Note that duality symmetry can also be broken in non-dispersive media where $\epsilon \neq \mu$.

The definition of optical helicity in the presence of material losses faces a set of challenges: For divergence-free displacement fields, the electric vector potential $\mathbf{C}$, inherent to the optical helicity, is written as $\mathbf{E} = -\nabla \times \mathbf{C}$ [49]. Gauss' law in the presence of sources and currents, $\nabla \cdot \mathbf{D} = \rho_0$ [64], elucidates two cases for which this condition is met (i) systems free of primary sources ($\rho_0 = 0$)

and (ii) lossless media. Systems obeying (i) and (ii) have been rigorously studied in previous work [50,68,71,82]—note that the model presented in [71] can be applied to systems with negligible material losses, such as perfect metals.

For applications in nanophotonics, we focus on chiral light–matter interactions in artificial nanostructures composed of linear, homogeneous, isotropic media, where material losses can play a significant role in the generation of chiral electromagnetic fields [66,67,83]. We consider time-harmonic electromagnetic fields with notation $\mathbf{E}(\mathbf{r},t) = \mathrm{Re}[\mathcal{E}(\mathbf{r})e^{-i\omega t}]$ for the electric field, where $\mathcal{E}(\mathbf{r})$ is the complex electric field amplitude of the electric field at spatial coordinate $\mathbf{r}$. For short-hand notation, we write complex field amplitudes as $\mathcal{E}(\mathbf{r}) = \mathcal{E}$. In linear media, the complex electric permittivity $\epsilon = \epsilon_0(\epsilon' + i\epsilon'')$ has its imaginary part of the relative permittivity $\epsilon'' = \sigma(\mathbf{r},\omega)/(\epsilon_0 \omega)$, where σ is the conductivity and $\mathcal{J}_{\mathrm{cond}} = \sigma(\mathbf{r},\omega)\mathcal{E}(\mathbf{r})$ is the complex amplitude of the conduction current density. We then reformulate Gauss' law as [61,64]:

$$\nabla \cdot \mathcal{D} = \nabla \cdot (\epsilon_0 \epsilon' \mathcal{E}) + \frac{i}{\omega} \underbrace{\nabla \cdot (\sigma(\mathbf{r},\omega)\mathcal{E})}_{\nabla \cdot \mathcal{J}_{\mathrm{cond}} = i\omega \rho_{\mathrm{cond}}(\mathbf{r})} = \rho_0(\mathbf{r}), \qquad (6)$$

with angular frequency ω. Equation (6) demonstrates that the electric displacement fields are not divergence free in lossy media. Specifically, the underbrace in Equation (6) shows time-harmonic charge continuity, from which Gauss' law can be reformulated as: $\nabla \cdot (\epsilon_0 \epsilon' \mathcal{E}) = \rho_0(\mathbf{r}) + \rho_{\mathrm{cond}}(\mathbf{r})$.

Thus, revisiting the case of an elastic and inelastic collision in classical mechanics (Figure 2), we draw a qualitative comparison to the physical relevance of optical helicity and optical chirality in the presence of matter. From Noether's theorem, the conservation of linear momentum in a lossless system arises from translational invariance [50,56] and, as Figure 2 demonstrates, the conservation of linear momentum captures the translational motion of a mechanical collision. Similarly, the conservation of optical helicity, with units of angular momentum, finds its physically relevant application in the description of chiral symmetries of propagating electromagnetic fields in systems where electromagnetic duality symmetry holds. In contrast, just as energy conservation captures the conversion of kinetic energy to heat in the case of an inelastic collision, optical chirality conservation is the suitable conservation law to describe the physical mechanism of optical chirality dissipation in the presence of lossy, dispersive media.

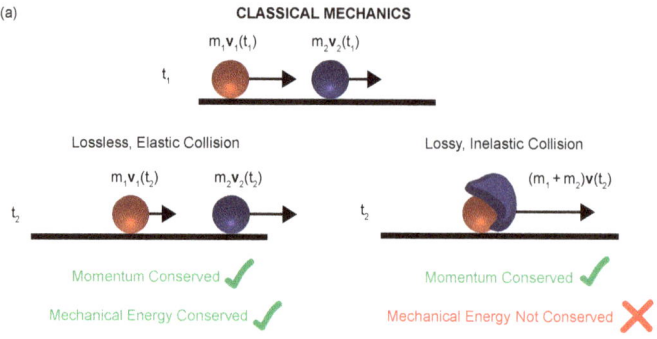

Figure 2. (**a**) Illustration of an elastic (**left**) and inelastic (**right**) collision of two objects with masses m_1, m_2, moving at velocities $\mathbf{v}_1$, $\mathbf{v}_2$ at times t_1, before the collision (**top**) and t_2, after the collision (**bottom**), respectively. While the total linear momentum of the system $\mathbf{p} = m\mathbf{v}$ is conserved for the elastic (**left**) and inelastic collision (**right**), the mechanical (kinetic) energy is not conserved for the inelastic collision due to energy dissipation. (**b**) Illustration of the interaction between chiral light and macroscopic matter for a lossless, dual-symmetric medium (**left**) and a lossy, dispersive medium (**right**). While both optical helicity and optical chirality conservation hold for the lossless, dual-symmetric case (**left**), the presence of a lossy, dispersive medium (**right**) breaks duality symmetry and helicity conservation no longer holds in its physically observable form. In contrast, the conservation law of optical chirality can be formulated to account for dissipative effects in the presence of lossy, dispersive media.

4.1. Observables Derived from Chiral Electromagnetism

In systems free of primary sources for time-harmonic fields, the conservation law of optical chirality (Equation (5)) in linear, dispersive media with losses is written as [66]:

$$-2\omega \int_V \text{Im}(\chi_e - \chi_m) d^3x + \int_V \text{Re}(\nabla \cdot \boldsymbol{\mathcal{F}}) d^3x = 0. \tag{7}$$

The first term in Equation (7) represents optical chirality dissipation:

$$\text{Im}(\chi_e - \chi_m) = \tag{8}$$
$$\frac{1}{8}[-\nabla \varepsilon' \cdot \text{Im}(\boldsymbol{\mathcal{E}} \times \boldsymbol{\mathcal{E}}^*) - \nabla \mu' \cdot \text{Im}(\boldsymbol{\mathcal{H}} \times \boldsymbol{\mathcal{H}}^*)]$$
$$+ \frac{1}{4}\omega(\varepsilon'\mu'' + \varepsilon''\mu')\text{Im}(\boldsymbol{\mathcal{E}}^* \cdot \boldsymbol{\mathcal{H}}),$$

where $\boldsymbol{\mathcal{E}}$ and $\boldsymbol{\mathcal{H}}$ are complex field amplitudes of the electric and magnetic induction fields, with $\epsilon = \epsilon' + i\epsilon''$ as the complex electric permittivity and $\mu = \mu' + i\mu''$ as the complex magnetic permeability.

Equation (8) is composed of two physically distinct terms, where the first represents optical chirality dissipation arising from material anisotropy, expressed by the gradient of ϵ' and μ', and the second describes optical chirality dissipation due to material loss, expressed by the imaginary parts of the material functions ϵ'' and μ'' [66,84]. The second term in Equation (7) represents the volume-integrated optical chirality flux, with density $\mathcal{F}$ defined as:

$$\mathcal{F} = \frac{1}{4}[\mathcal{E} \times (\nabla \times \mathcal{H}^*) - \mathcal{H}^* \times (\nabla \times \mathcal{E})]. \tag{9}$$

In the far field, where electromagnetic fields are well-approximated as plane waves, $\mathcal{F}$ can be represented by a weighted superposition of the optical chirality flux density arising from left- ($\mathcal{F}_{LCPL}$) and right-handed ($\mathcal{F}_{RCPL}$) circularly polarized plane waves, yielding a total far-field (FF) optical chirality flux $\mathcal{F}_{FF} = |l|^2 \mathcal{F}_{LCPL} + |r|^2 \mathcal{F}_{RCPL}$. The weighting factors for left- and right-handed CPL are represented by constants l and r, respectively. From this, $\mathcal{F}_{FF}$ is directly proportional to the third Stokes parameter S_3, describing the degree of circular polarization, as [66,85]:

$$\mathcal{F}_{FF} = \frac{\omega}{c}(|l|^2 \mathcal{S}_{LCPL} - |r|^2 \mathcal{S}_{RCPL}) \propto S_3, \tag{10}$$

where c is the speed of light and $\mathcal{S} = \mathcal{E} \times \mathcal{H}^*$ is the complex amplitude of the Poynting vector [64]. The optical chirality flux generated by lossy, dispersive media has been experimentally observed in the far field for periodic arrays of two-dimensionally chiral metallic nanoantennas [83] and colloidal dispersions of three-dimensionally chiral metallic nanopyramids at the single-particle level [86]. These experimental results demonstrated how the optical chirality flux is a physically relevant far-field observable, with the ability to provide information on chiral light–matter interactions in the near and far field.

In addition to the physically observable description of optical chirality dissipation and flux, derived from the conservation law of optical chirality (Equation (7)), the optical chirality density χ (Table 1) has the ability to locally quantify the chirality of electromagnetic fields which can be strongly enhanced upon interaction with matter. The free-space optical chirality density (Table 1) of time-averaged, time-harmonic fields is written as:

$$\bar{\chi} = -\frac{\omega}{2c^2}\text{Im}(\mathcal{E}^* \cdot \mathcal{H}) = -\frac{\omega}{2c^2}|\mathcal{E}||\mathcal{H}|\cos(\beta_{i\mathcal{E},\mathcal{H}}), \tag{11}$$

where $\mathcal{E}$ and $\mathcal{H}$ are the complex electric field and magnetic field amplitudes, $\beta_{i\mathcal{E},\mathcal{H}}$ is the angle between the product $i\mathcal{E}$ and the $\mathcal{H}$ field, and the overbar in $\bar{\chi}$ denotes the time average. Tang and Cohen identified Equation (11) within the excitation rate equation for chiral molecules [54]. This finding revealed the possibility to increase chiral selectivity of molecular excitation by orders of magnitude, when chiral molecules interact with electromagnetic fields of enhanced $\bar{\chi}$ [6]. Subsequently, the rapidly-developing research area of chiral nanophotonics, summarized in a series of review articles [25,40–42], devoted itself to constructing solutions to Maxwell's equations for which $\bar{\chi}$ exceeds its corresponding value for CPL, where $\bar{\chi}/|\bar{\chi}_{CPL}|$ was termed *optical chirality enhancement* by Schäferling et al. [87].

A single electromagnetic plane wave reaches its maximum value of $\bar{\chi}$ at the circular polarization state, resulting in $|\bar{\chi}|/|\bar{\chi}_{CPL}| = 1$ [83]. However, the scale discrepancy between chiral molecules and the wavelength of CPL results in inherently weak selectivity [54]. Optical chirality enhancement beyond unity coincides with the concentration of electromagnetic energy ($w = w_e + w_m$), as $\bar{\chi}$ is bounded by $c|\bar{\chi}|/(\omega w) \leq 1$ for speed of light c and angular frequency ω [88,89].

In free space, theoretical studies have predicted $|\bar{\chi}|/|\bar{\chi}_{CPL}| > 1$ for the diffraction-limited focusing of a circularly polarized Gaussian beam or the appropriate superposition of two Gaussian beams with radial and azimuthal polarization [90,91]. To better match molecular dimensions, evanescent waves can achieve a theoretically-unlimited spatial concentration of electromagnetic fields at material interfaces [92], enabling optical chirality enhancement beyond the diffraction limit [83]. In particular, artificial nanostructures, with dimensions comparable to the wavelength of light, show great promise for the rational design of concentrated chiral electromagnetic fields.

We now provide a physical interpretation on how the optical chirality density χ and optical chirality flux density $\mathcal{F}$ of electromagnetic fields (Table 1) can be enhanced upon interaction with matter. A qualitative comparison can be made to vortex flow in fluid dynamics [61]. From Equation (1), the helicity density of a fluid vortex is written as $h_\text{fluid} = \mathbf{v} \cdot (\nabla \times \mathbf{v})$ for fluid velocity $\mathbf{v}$ and vorticity $\omega = \nabla \times \mathbf{v}$ [47]. Further, the flux of the solenoidal vorticity ω is conserved, $\nabla \cdot \omega = 0$, indicating that an equal number of vortices with clockwise or counterclockwise rotation crosses the boundary of a closed system [47].

The formal analogy between the $\mathbf{v}$ field of h_fluid and the electric and magnetic fields of χ allows for the interpretation of χ as the wrapping density of electric and magnetic field lines around their rotation axis [54]. With increasing wrapping density along this axis, χ will increase in value. Further, while they differ in form, the vorticity flux $\nabla \cdot \omega = 0$ can support the interpretation of the optical chirality flux $\nabla \cdot \mathcal{F}$. In particular, $\nabla \cdot \mathcal{F} \neq 0$ when the boundary of a studied system is traversed by an excess of one handedness of chiral electromagnetic field lines.

In contrast to fluidics, the interplay between electric and magnetic fields gives rise to additional complexity in electromagnetic chirality. In particular, Maxwell's equations dictate that a magnetic field can be induced by the rotation of an electric field and vice versa [64]. Further, while solid interfaces can act as sources of fluid vorticity, affecting fluid flow on boundary layers [93], the conservation law of optical chirality (Equation (7)) elucidates how material charges and currents can act as sources or sinks of chiral electromagnetic fields. In particular, both material loss and anisotropy (Equation (8)) can result in the dissipation and generation of electromagnetic chirality [54,66,67,70,83].

5. Chiral Light–Matter Interactions in Artificial Nanostructures

The effective design and optimization of artificial nanostructures with respect to their chiral optical fields requires insight into the mechanism of their interaction with chiral light. This section discusses the ability of metallic and dielectric nanostructures to generate chiral electromagnetic fields and elucidates the physical mechanisms present in each case. Figure 3 shows numerical simulations of the optical chirality flux generated by gold (part a) and silicon (part b) nanospheres of 75 nm radius (COMSOL Multiphysics 5.3a, gold material functions from Johnson and Christy [94] and silicon material functions from Aspnes and Studna [95]). In both systems, linearly polarized plane-wave excitation (LP, black) does not generate an optical chirality flux. In contrast, excitation with left- (LCP, red) and right-handed (RCP, blue) circularly polarized plane waves results in mirror-symmetric optical chirality flux spectra. Thus, in an achiral system, such as the studied nanospheres, inversion of the sign of the excitation light source inverts the sign of the optical chirality flux generated by the nanostructure.

The conservation law of optical chirality (Equation (7)), which sets equal the physical mechanisms of optical chirality dissipation and optical chirality flux can explain the generation of chiral optical fields in Figure 3. In particular, a non-zero optical chirality dissipation arises from the interaction between the achiral nanospheres and the chiral excitation source, leading to the generation of an optical chirality flux. In contrast, prior work has shown that nanostructures with a chiral geometry interacting with achiral, linearly polarized light have the ability to dissipate optical chirality, thus generating an optical chirality flux [66,83,86,96]. Figure 3 also demonstrates that the silicon nanosphere generates an optical chirality flux an order of magnitude larger than the gold nanosphere of the same size.

Beyond the simple case of spherical nanoparticles shown in Figure 3, we now discuss distinct mechanisms which can contribute to enhancement of the optical chirality flux generated by metallic and dielectric nanostructures. The delocalized surface-electron gas, oscillating on resonance in metallic nanostructures [97], results in polarization and conduction currents which can interact in a sensitive and selective manner with chiral electromagnetic fields [61]. Thus, metallic nanoparticles exhibiting a left- or right-handed chiral geometry can effectively dissipate optical chirality and generate an optical chirality flux [66,83,86,96]. A myriad of research efforts have, therefore, realized metallic nanostructures with complex chiral geometries, such as metallic helices, pyramids, dimers, and oligomers [10–28] (see also review articles [25,39–41]).

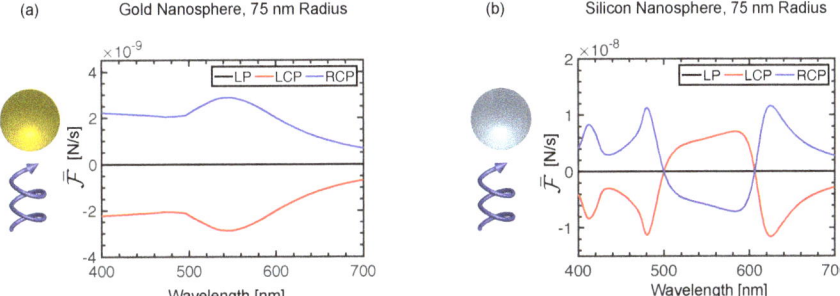

Figure 3. (**a**) Schematic illustration of a gold nanoparticle (spherical geometry, 75 nm radius) interacting with a circularly polarized plane wave. Numerical simulations of the total, volume-integrated optical chirality flux $\bar{\mathcal{F}} = \int_V \nabla \cdot \mathcal{F} d^3 x$ of the gold nanosphere upon excitation with linearly polarized (LP, black), left-handed circularly polarized light (CPL) (LCP, red), and right-handed CPL (RCP, blue). (**b**) Schematic illustration of a silicon nanoparticle (spherical geometry, 75 nm radius) interacting with a circularly polarized plane wave. Numerical simulations of the total, volume-integrated optical chirality flux $\bar{\mathcal{F}}$ of the silicon nanosphere upon excitation with linearly polarized (LP, black), left-handed CPL (LCP, red), and right-handed CPL (RCP, blue).

In dielectric nanostructures, the conduction and polarization currents are considerably smaller than in metals [98]. Further, in the absence of primary sources, an intrinsic magnetic dipole moment arises from the magnetization current. Thus, tailoring the magnitude and phase shift of the intrinsic electric and magnetic dipole moments can enhance the chiral electromagnetic fields generated by dielectric nanostructures. This can be controlled by phase-shifted electric and magnetic fields in the excitation source, as is the case for CPL (Figure 3b), or further geometric tuning, as was demonstrated in recent research for achiral silicon nanospheres [29,30], silicon disk and sphere metasurfaces [32,33,35,36], or dielectric dimer structures [31,34,37,38]. These additional degrees of freedom inherent to the mechanism of chiral light–matter interactions in dielectric nanostructures enable the generation of highly enhanced chiral electromagnetic fields in simplified geometric configurations, suitable for high-throughput applications where strong optical chirality enhancement can be rationally designed in the nanostructure near field.

6. Conclusions

In conclusion, this perspective provides insight on the physical applicability of the optical helicity and the optical chirality in free space and in the presence of matter. In free space, a qualitative parallel between momentum in classical mechanics and optical helicity in classical electrodynamics can be made; likewise, a parallel between force and optical chirality also exists. We applied time and parity symmetry relations to demonstrate how the optical chirality density and flux quantify the handedness of an electromagnetic field. When chiral light interacts with macroscopic matter, we then identified how the optical helicity provides useful physical information for the case of lossless, dual-symmetric media, while the optical chirality provides physically observable information in the case of lossy, dispersive media. Here, a comparison to energy and momentum conservation for the lossless, elastic collision and the lossy, inelastic collision of two moving objects provides insight on the applicability of optical helicity and optical chirality conservation in the presence of matter. Finally, we applied the conservation law of optical chirality to numerically simulate the optical chirality flux generated by a gold and silicon nanosphere of 75 nm radius. While no optical chirality flux was generated upon linearly polarized excitation, left- and right-handed CPL resulted in mirror-symmetric optical chirality flux spectra in both cases. This effect can be further enhanced by tuning the geometry of the nanostructure; while metallic nanostructures with a chiral shape direct the currents arising from the surface-electron gas, the interplay between electric and magnetic dipole moments in dielectric nanostructures affects the generation of chiral light. This information provides a platform from which researchers can improve the rational design of nanophotonic structures for the optimized enhancement of chiral light–matter interactions.

Author Contributions: Conceptualization, L.V.P., J.A.D. and A.G.-E.; methodology, L.V.P., J.A.D. and A.G.-E.; software, L.V.P.; validation, L.V.P., J.A.D. and A.G.-E.; formal analysis, L.V.P. and A.G.-E.; investigation, L.V.P., J.A.D. and A.G.-E.; resources, L.V.P., J.A.D. and A.G.-E.; data curation, L.V.P.; writing—original draft preparation, L.V.P.; writing—review and editing, L.V.P., J.A.D. and A.G.-E.; visualization, L.V.P.; supervision, J.A.D. and A.G.-E.; project administration, J.A.D. and A.G.-E.; funding acquisition, L.V.P., J.A.D. and A.G.-E.

Funding: This research was supported by the Swiss National Science Foundation Early Postdoc.Mobility Fellowship, project number P2EZP2_181595, Eusko Jaurlaritza, grant numbers PI-2016-1-0041, KK-2017/00089, IT1164-19 and KK-2019/00101, Ministerio de Economia, Industria y Competitividad, Gobierno de Espana grant number FIS2016-80174-P. A.G.-E. was funded by the Fellows Gipuzkoa fellowship of the Gipuzkoako Foru Aldundia through FEDER "Una Manera de hacer Europa".

Acknowledgments: The authors thank David J. Norris, Lukas Novotny, Christian Hafner, Mark Lawrence, John Abendroth, Michelle Solomon, Jack Hu, David R. Barton III, Elissa Klopfer and Shing-Shing Ho for helpful scientific feedback. We thank Shing-Shing Ho for contributions to artistic rendering.

Conflicts of Interest: The authors declare no conflict of interest.

References

1. Richardson, G. *The Foundations of Stereo Chemistry: Memoirs by Pasteur, Van't Hoff, Lebel and Wislicenus*; American Book Company: New York, NY, USA, 1901.
2. Eriksson, T.; Björkman, S.; Höglund, P. Clinical pharmacology of thalidomide. *Eu. J. Clin. Pharmacol.* **2001**, *57*, 365–376. [CrossRef] [PubMed]
3. Seo, J.S.; Whang, D.; Lee, H.; Im Jun, S.; Oh, J.; Jeon, Y.J.; Kim, K. A homochiral metal-organic porous material for enantioselective separation and catalysis. *Nature* **2000**, *404*, 982. [CrossRef] [PubMed]
4. Ma, L.; Abney, C.; Lin, W. Enantioselective catalysis with homochiral metal-organic frameworks. *Chem. Soc. Rev.* **2009**, *38*, 1248–1256. [CrossRef] [PubMed]
5. Blaser, H.U.; Federsel, H.J. *Asymmetric Catalysis on Industrial Scale: Challenges, Approaches and Solutions*; Wiley-VCH Verlag GmbH & Co. KGaA: Weinheim, Germany , 2004.
6. Tang, Y.; Cohen, A.E. Enhanced enantioselectivity in excitation of chiral molecules by superchiral light. *Science* **2011**, *332*, 333–336. [CrossRef] [PubMed]
7. Zhao, Y.; Saleh, A.A.; van de Haar, M.A.; Baum, B.; Briggs, J.A.; Lay, A.; Reyes-Becerra, O.A.; Dionne, J.A. Nanoscopic control and quantification of enantioselective optical forces. *Nat. Nanotechnol.* **2017**, *12*, 1055. [CrossRef] [PubMed]
8. Petersen, J.; Volz, J.; Rauschenbeutel, A. Chiral nanophotonic waveguide interface based on spin-orbit interaction of light. *Science* **2014**, *346*, 67–71. [CrossRef]
9. Le Feber, B.; Rotenberg, N.; Kuipers, L. Nanophotonic control of circular dipole emission. *Nat. Commun.* **2015**, *6*, 6695. [CrossRef]
10. Papakostas, A.; Potts, A.; Bagnall, D.; Prosvirnin, S.; Coles, H.; Zheludev, N. Optical manifestations of planar chirality. *Phys. Rev. Lett.* **2003**, *90*, 107404. [CrossRef]
11. Gansel, J.K.; Thiel, M.; Rill, M.S.; Decker, M.; Bade, K.; Saile, V.; von Freymann, G.; Linden, S.; Wegener, M. Gold helix photonic metamaterial as broadband circular polarizer. *Science* **2009**, *325*, 1513–1515. [CrossRef]
12. Hendry, E.; Carpy, T.; Johnston, J.; Popland, M.; Mikhaylovskiy, R.; Lapthorn, A.; Kelly, S.; Barron, L.; Gadegaard, N.; Kadodwala, M. Ultrasensitive detection and characterization of biomolecules using superchiral fields. *Nat. Nanotechnol.* **2010**, *5*, 783–787. [CrossRef]
13. Hentschel, M.; Schäferling, M.; Weiss, T.; Liu, N.; Giessen, H. Three-dimensional chiral plasmonic oligomers. *Nano Lett.* **2012**, *12*, 2542–2547. [CrossRef] [PubMed]

14. Zhao, Y.; Belkin, M.; Alù, A. Twisted optical metamaterials for planarized ultrathin broadband circular polarizers. *Nat. Commun.* **2012**, *3*, 870. [CrossRef] [PubMed]
15. Yin, X.; Schäferling, M.; Metzger, B.; Giessen, H. Interpreting chiral nanophotonic spectra: The plasmonic Born-Kuhn model. *Nano Lett.* **2013**, *13*, 6238–6243. [CrossRef] [PubMed]
16. Mark, A.G.; Gibbs, J.G.; Lee, T.C.; Fischer, P. Hybrid nanocolloids with programmed three-dimensional shape and material composition. *Nat. Mater.* **2013**, *12*, 802. [CrossRef] [PubMed]
17. Schamel, D.; Pfeifer, M.; Gibbs, J.G.; Miksch, B.r.; Mark, A.G.; Fischer, P. Chiral Colloidal Molecules and Observation of the Propeller Effect. *J. Am. Chem. Soc.* **2013**, *135*, 12353–12359. [CrossRef] [PubMed]
18. Schäferling, M.; Yin, X.; Engheta, N.; Giessen, H. Helical plasmonic nanostructures as prototypical chiral near-field sources. *ACS Photonics* **2014**, *1*, 530–537. [CrossRef]
19. McPeak, K.M.; van Engers, C.D.; Blome, M.; Park, J.H.; Burger, S.; Gosalvez, M.A.; Faridi, A.; Ries, Y.R.; Sahu, A.; Norris, D.J. Complex Chiral Colloids and Surfaces via High-Index Off-Cut Silicon. *Nano Lett.* **2014**, *14*, 2934–2940. [CrossRef]
20. Yeom, J.; Yeom, B.; Chan, H.; Smith, K.W.; Dominguez-Medina, S.; Bahng, J.H.; Zhao, G.; Chang, W.S.; Chang, S.J.; Chuvilin, A.; et al. Chiral templating of self-assembling nanostructures by circularly polarized light. *Nat. Mater.* **2015**, *14*, 66. [CrossRef]
21. Wang, L.Y.; Smith, K.W.; Dominguez-Medina, S.; Moody, N.; Olson, J.M.; Zhang, H.; Chang, W.S.; Kotov, N.; Link, S. Circular Differential Scattering of Single Chiral Self-Assembled Gold Nanorod Dimers. *ACS Photonics* **2015**, *2*, 1602–1610. [CrossRef]
22. McPeak, K.M.; van Engers, C.D.; Bianchi, S.; Rossinelli, A.; Poulikakos, L.V.; Bernard, L.; Herrmann, S.; Kim, D.K.; Burger, S.; Blome, M.; et al. Ultraviolet plasmonic chirality from colloidal aluminum nanoparticles exhibiting charge-selective protein detection. *Adv. Mater.* **2015**, *27*, 6244–6250. [CrossRef]
23. Tullius, R.; Karimullah, A.S.; Rodier, M.; Fitzpatrick, B.; Gadegaard, N.; Barron, L.D.; Rotello, V.M.; Cooke, G.; Lapthorn, A.; Kadodwala, M. "Superchiral" spectroscopy: Detection of protein higher order hierarchical structure with chiral plasmonic nanostructures. *J. Am. Chem. Soc.* **2015**, *137*, 8380–8383. [CrossRef] [PubMed]
24. Kosters, D.; De Hoogh, A.; Zeijlemaker, H.; Acar, H.; Rotenberg, N.; Kuipers, L. Core-shell plasmonic nanohelices. *ACS Photonics* **2017**, *4*, 1858–1863. [CrossRef] [PubMed]
25. Hentschel, M.; Schäferling, M.; Duan, X.; Giessen, H.; Liu, N. Chiral plasmonics. *Sci. Adv.* **2017**, *3*, e1602735. [CrossRef] [PubMed]
26. Karst, J.; Strohfeldt, N.; Schäferling, M.; Giessen, H.; Hentschel, M. Single plasmonic oligomer chiral spectroscopy. *Adv. Opt. Mater.* **2018**, *6*, 1800087. [CrossRef]
27. Lee, H.E.; Ahn, H.Y.; Mun, J.; Lee, Y.Y.; Kim, M.; Cho, N.H.; Chang, K.; Kim, W.S.; Rho, J.; Nam, K.T. Amino-acid-and peptide-directed synthesis of chiral plasmonic gold nanoparticles. *Nature* **2018**, *556*, 360. [CrossRef] [PubMed]
28. Karst, J.; Cho, N.H.; Kim, H.; Lee, H.E.; Nam, K.T.; Giessen, H.; Hentschel, M. Chiral scatterometry on chemically synthesized single plasmonic nanoparticles. *ACS Nano* **2019**. [CrossRef] [PubMed]
29. García-Etxarri, A.; Dionne, J.A. Surface-enhanced circular dichroism spectroscopy mediated by nonchiral nanoantennas. *Phys. Rev. B* **2013**, *87*, 235409. [CrossRef]
30. Ho, C.S.; Garcia-Etxarri, A.; Zhao, Y.; Dionne, J. Enhancing enantioselective absorption using dielectric nanospheres. *ACS Photonics* **2017**, *4*, 197–203. [CrossRef]
31. Zhang, W.; Wu, T.; Wang, R.; Zhang, X. Amplification of the molecular chiroptical effect by low-loss dielectric nanoantennas. *Nanoscale* **2017**, *9*, 5701–5707. [CrossRef] [PubMed]
32. Mohammadi, E.; Tsakmakidis, K.L.; Askarpour, A.N.; Dehkhoda, P.; Tavakoli, A.; Altug, H. Nanophotonic platforms for enhanced chiral sensing. *ACS Photonics* **2018**, *5*, 2669–2675. [CrossRef]
33. Solomon, M.L.; Hu, J.; Lawrence, M.; García-Etxarri, A.; Dionne, J.A. Enantiospecific optical enhancement of chiral sensing and separation with dielectric metasurfaces. *ACS Photonics* **2018**, *6*, 43–49. [CrossRef]
34. Yao, K.; Liu, Y. Enhancing circular dichroism by chiral hotspots in silicon nanocube dimers. *Nanoscale* **2018**, *10*, 8779–8786. [CrossRef] [PubMed]
35. Graf, F.; Feis, J.; Garcia-Santiago, X.; Wegener, M.; Rockstuhl, C.; Fernandez-Corbaton, I. Achiral, helicity preserving, and resonant structures for enhanced sensing of chiral molecules. *ACS Photonics* **2019**, *6*, 482–491. [CrossRef]
36. Hanifeh, M.; Capolino, F. Helicity density enhancement in a planar array of achiral high-density dielectric nanoparticles. *arXiv* **2019**, arXiv:1905.03387.

37. Zhao, X.; Reinhard, B.M. Switchable Chiroptical Hot-Spots in Silicon Nanodisk Dimers. *ACS Photonics* **2019**. [CrossRef]
38. Mohammadi, E.; Tavakoli, A.; Dehkhoda, P.; Jahani, Y.; Tsakmakidis, K.L.; Tittl, A.; Altug, H. Accessible superchiral near-fields driven by tailored electric and magnetic resonances in all-dielectric nanostructures. *ACS Photonics* **2019**. [CrossRef]
39. Govorov, A.O.; Gun'ko, Y.K.; Slocik, J.M.; Gérard, V.A.; Fan, Z.; Naik, R.R. Chiral nanoparticle assemblies: Circular dichroism, plasmonic interactions, and exciton effects. *J. Mater. Chem.* **2011**, *21*, 16806–16818. [CrossRef]
40. Valev, V.K.; Baumberg, J.J.; Sibilia, C.; Verbiest, T. Chirality and chiroptical effects in plasmonic nanostructures: fundamentals, recent progress, and outlook. *Adv. Mater.* **2013**, *25*, 2517–2534. [CrossRef]
41. Ben-Moshe, A.; Maoz, B.M.; Govorov, A.O.; Markovich, G. Chirality and chiroptical effects in inorganic nanocrystal systems with plasmon and exciton resonances. *Chem. Soc. Rev.* **2013**, *42*, 7028–7041. [CrossRef]
42. Smith, K.W.; Link, S.; Chang, W.S. Optical characterization of chiral plasmonic nanostructures. *J. Photochem. Photobiol. C Photochem. Rev.* **2017**, *32*, 40–57. [CrossRef]
43. Barnett, S.M.; Cameron, R.P.; Yao, A.M. Duplex symmetry and its relation to the conservation of optical helicity. *Phys. Rev. A* **2012**, *86*, 013845. [CrossRef]
44. Crimin, F.; Mackinnon, N.; Götte, J.; Barnett, S. Optical helicity and chirality: Conservation and sources. *Appl. Sci.* **2019**, *9*, 828. [CrossRef]
45. Tung, W.K. *Group Theory in Physics: An Introduction to Symmetry Principles, Group Representations, and Special Functions in Classical and Quantum Physics*; World Scientific Publishing Company: Singapore, 1985.
46. Schwartz, M.D. *Quantum Field Theory and the Standard Model*; Cambridge University Press: Cambridge, UK, 2014.
47. Moffatt, H.K. The degree of knottedness of tangled vortex lines. *J. Fluid Mech.* **1969**, *35*, 117–129. [CrossRef]
48. Berger, M.A. Introduction to magnetic helicity. *Plasma Phys. Control. Fusion* **1999**, *41*, B167. [CrossRef]
49. Cameron, R.P. On the "second potential" in electrodynamics. *J. Opt.* **2013**, *16*, 015708. [CrossRef]
50. Fernandez-Corbaton, I.; Zambrana-Puyalto, X.; Tischler, N.; Vidal, X.; Juan, M.L.; Molina-Terriza, G. Electromagnetic duality symmetry and helicity conservation for the macroscopic maxwells equations. *Phys. Rev. Lett.* **2013**, *111*, 060401. [CrossRef] [PubMed]
51. Cameron, R.P.; Barnett, S.M.; Yao, A.M. Optical helicity, optical spin and related quantities in electromagnetic theory. *New J. Phys.* **2012**, *14*, 053050. [CrossRef]
52. Candlin, D. Analysis of the New Conservation Law in Electromagnetic Theory. *Il Nuovo Cimento (1955–1965)* **1965**, *37*, 1390–1395. [CrossRef]
53. Kibble, T. Conservation laws for free fields. *J. Math. Phys.* **1965**, *6*, 1022–1026. [CrossRef]
54. Tang, Y.; Cohen, A.E. Optical chirality and its interaction with matter. *Phys. Rev. Lett.* **2010**, *104*, 163901. [CrossRef]
55. Lipkin, D.M. Existence of a New Conservation Law in Electromagnetic Theory. *J. Math. Phys.* **1964**, *5*, 696–700. [CrossRef]
56. Noether, E. Invariante variations probleme. *Nachrichten von der Königlichen Gesellschaft der Wissenschaften zu Gottingen (Royal Society of Sciences, Gottingen)* **1918**, *235–257*, 1918.
57. Calkin, M.G. An Invariance Property of the Free Electromagnetic Field. *Am. J. Phys.* **1965**, *33*, 958–960. [CrossRef]
58. Bliokh, K.Y.; Bekshaev, A.Y.; Nori, F. Dual electromagnetism: Helicity, spin, momentum and angular momentum. *New J. Phys.* **2013**, *15*, 033026. [CrossRef]
59. Sayir, M.; Dual, J.; Kaufmann, S. *Ingenieurmechanik 1*; Springer: Wiesbaden, Germany, 2008.
60. Hafner, C. (ETH Zürich, Zürich, ZH, Switzerland). Personal Communication, 2016.
61. Poulikakos, L.V. Chiral Light–Matter Interactions in the Near and Far Field. Ph.D. Thesis, ETH Zürich, Zürich, Switzerland, 2018.
62. Guasti, M.F. Chirality, helicity and the rotational content of electromagnetic fields. *Phys. Lett. A* **2019**, *383*, 3180–3186. [CrossRef]
63. Barron, L. *Chirality at the Nanoscale*; Wiley-VCH Verlag: Weinheim, BW, Germany; GmbH and Co. KGaA: Dusseldorf, Germany, 2009.
64. Jackson, J.D. *Classical Electrodynamics*, 3rd ed.; John Wiley & Sons, Inc.: New York, NY, USA, 1999.

65. Hafner, C. *Numerische Berechnung Elektromagnetischer Felder, Grundlagen, Methoden, Anwendungen*; Springer: Berlin/Heidelberg, Germany, 1987.
66. Poulikakos, L.V.; Gutsche, P.; McPeak, K.M.; Burger, S.; Niegemann, J.; Hafner, C.; Norris, D.J. Optical chirality flux as a useful far-field probe of chiral near fields. *ACS Photonics* **2016**, *3*, 1619–1625. [CrossRef]
67. Vázquez-Lozano, J.E.; Martínez, A. Optical chirality in dispersive and lossy media. *Phys. Rev. Lett.* **2018**, *121*, 043901. [CrossRef]
68. Philbin, T.G. Lipkin's conservation law, noether's theorem, and the relation to optical helicity. *Phys. Rev. A* **2013**, *87*, 043843. [CrossRef]
69. Van Kruining, K.; Götte, J.B. The conditions for the preservation of duality symmetry in a linear medium. *J. Opt.* **2016**, *18*, 085601. [CrossRef]
70. Nienhuis, G. Conservation laws and symmetry transformations of the electromagnetic field with sources. *Phys. Rev. A* **2016**, *93*, 023840. [CrossRef]
71. Alpeggiani, F.; Bliokh, K.; Nori, F.; Kuipers, L. Electromagnetic helicity in complex media. *Phys. Rev. Lett.* **2018**, *120*, 243605. [CrossRef] [PubMed]
72. Zambrana-Puyalto, X.; Vidal, X.; Juan, M.L.; Molina-Terriza, G. Dual and anti-dual modes in dielectric spheres. *Opt. Express* **2013**, *21*, 17520–17530. [CrossRef] [PubMed]
73. Zambrana-Puyalto, X.; Fernandez-Corbaton, I.; Juan, M.; Vidal, X.; Molina-Terriza, G. Duality symmetry and Kerker conditions. *Opt. Lett.* **2013**, *38*, 1857–1859. [CrossRef] [PubMed]
74. Schmidt, M.K.; Aizpurua, J.; Zambrana-Puyalto, X.; Vidal, X.; Molina-Terriza, G.; Sáenz, J.J. Isotropically polarized speckle patterns. *Phys. Rev. Lett.* **2015**, *114*, 113902. [CrossRef] [PubMed]
75. Allen, L.; Beijersbergen, M.W.; Spreeuw, R.; Woerdman, J. Orbital angular momentum of light and the transformation of Laguerre-Gaussian laser modes. *Phys. Rev. A* **1992**, *45*, 8185. [CrossRef] [PubMed]
76. Simpson, N.; Dholakia, K.; Allen, L.; Padgett, M. Mechanical equivalence of spin and orbital angular momentum of light: An optical spanner. *Opt. Lett.* **1997**, *22*, 52–54. [CrossRef]
77. Padgett, M.; Courtial, J.; Allen, L. Light's orbital angular momentum. *Phys. Today* **2004**, *57*, 35–40. [CrossRef]
78. Garcia-Etxarri, A. Optical polarization mobius strips on all-dielectric optical scatterers. *ACS Photonics* **2017**, *4*, 1159–1164. [CrossRef]
79. Olmos-Trigo, J.; Sanz-Fernández, C.; García-Etxarri, A.; Molina-Terriza, G.; Bergeret, F.S.; Sáenz, J.J. Enhanced spin-orbit optical mirages from dual nanospheres. *Phys. Rev. A* **2019**, *99*, 013852. [CrossRef]
80. Olmos-Trigo, J.; Sanz-Fernández, C.; Abujetas, D.R.; García-Etxarri, A.; Molina-Terriza, G.; Sánchez-Gil, J.; Bergeret, S.F.; Saénz, J.J. Role of the absorption on the spin-orbit interactions of light with Si nano-particles. *arXiv* **2019**, arXiv:1903.03816.
81. Fernandez-Corbaton, I.; Zambrana-Puyalto, X.; Molina-Terriza, G. Helicity and angular momentum: A symmetry-based framework for the study of light-matter interactions. *Phys. Rev. A* **2012**, *86*, 042103. [CrossRef]
82. Nieto-Vesperinas, M. Optical theorem for the conservation of electromagnetic helicity: Significance for molecular energy transfer and enantiomeric discrimination by circular dichroism. *Phys. Rev. A* **2015**, *92*, 023813. [CrossRef]
83. Poulikakos, L.V.; Thureja, P.; Stollmann, A.; De Leo, E.; Norris, D.J. Chiral light design and detection inspired by optical antenna theory. *Nano Lett.* **2018**, *18*, 4633–4640. [CrossRef] [PubMed]
84. Gutsche, P.; Poulikakos, L.V.; Hammerschmidt, M.; Burger, S.; Schmidt, F. Time-harmonic optical chirality in inhomogeneous space. In *Opto, Photonic and Phononic Properties of Engineered Nanostructures VI*; International Society for Optics and Photonics: San Francisco, CA, USA, 2016; Volume 9756, p. 97560X.
85. Collett, E. *Field Guide to Polarization*; SPIE Press: Bellingham, WA, USA, 2005; Volume FG05.
86. Schnoering, G.; Poulikakos, L.V.; Rosales-Cabara, Y.; Canaguier-Durand, A.; Norris, D.J.; Genet, C. Three-dimensional enantiomeric recognition of optically trapped single chiral nanoparticles. *Phys. Rev. Lett.* **2018**, *121*, 023902. [CrossRef] [PubMed]
87. Schäferling, M.; Dregely, D.; Hentschel, M.; Giessen, H. Tailoring enhanced optical chirality: Design principles for chiral plasmonic nanostructures. *Phys. Rev. X* **2012**, *2*, 031010. [CrossRef]
88. Choi, J.S.; Cho, M. Limitations of a Superchiral Field. *Phys. Rev. A* **2012**, *86*, 063834. [CrossRef]
89. Canaguier-Durand, A.; Genet, C. Chiral near fields generated from plasmonic optical lattices. *Phys. Rev. A* **2014**, *90*, 023842. [CrossRef]

90. Hanifeh, M.; Albooyeh, M.; Capolino, F. Helicity maximization of structured light to empower nanoscale chiral matter interaction. *arXiv* **2018**, arXiv:1809.04119.
91. Hanifeh, M.; Albooyeh, M.; Capolino, F. Empowering Structured Light to Enhance Chirality Detection and Characterization at Nanoscale. In *Opto, Complex Light and Optical Forces XIII*; International Society for Optics and Photonics: San Francisco, CA, USA, 2019; ; Volume 10935, p. 1093504.
92. Novotny, L.; Hecht, B. *Principles of Nano-Optics*, 2nd ed.; Cambridge University Press: New York, NY, USA, 2012.
93. Moffatt, H.K. Helicity and Singular Structures in Fluid Dynamics. *Proc. Natl. Acad. Sci. USA* **2014**, *111*, 3663–3670. [CrossRef] [PubMed]
94. Johnson, P.B.; Christy, R.W. Optical Constants of the Noble Metals. *Phys. Rev. B* **1972**, *6*, 4370. [CrossRef]
95. Aspnes, D.E.; Studna, A.A. Dielectric functions and optical parameters of Si, Ge, GaP, GaAs, GaSb, InP, InAs, and InSb from 1.5 to 6.0 Ev. *Phys. Rev. B* **1983**, *27*, 985. [CrossRef]
96. Hashiyada, S.; Narushima, T.; Okamoto, H. Imaging chirality of optical fields near achiral metal nanostructures excited with linearly polarized light. *ACS Photonics* **2018**, *5*, 1486–1492. [CrossRef]
97. Maier, S.A. *Plasmonics: Fundamentals and Applications*; Springer Science & Business Media: New York, NY, USA, 2007.
98. Capolino, F. *Theory and Phenomena of Metamaterials*; CRC Press/Taylor & Francis: Boca Raton, FL, USA, 2009.

 © 2019 by the authors. Licensee MDPI, Basel, Switzerland. This article is an open access article distributed under the terms and conditions of the Creative Commons Attribution (CC BY) license (http://creativecommons.org/licenses/by/4.0/).

MDPI\
St. Alban-Anlage 66\
4052 Basel\
Switzerland\
Tel. +41 61 683 77 34\
Fax +41 61 302 89 18\
www.mdpi.com

Symmetry Editorial Office\
E-mail: symmetry@mdpi.com\
www.mdpi.com/journal/symmetry

www.ingramcontent.com/pod-product-compliance
Lightning Source LLC
LaVergne TN
LVHW070604100526
838202LV00012B/560